Risk and Financial Management in Construction

Risk and Financial Management in Construction

SIMON A. BURTONSHAW-GUNN

GOWER

Published by
Gower Publishing Limited
Wey Court East
Union Road
Farnham
Surrey GU9 7PT
England

Ashgate Publishing Company
Suite 420
101 Cherry Street
Burlington, VT 05401-4405
USA

Simon A. Burtonshaw-Gunn has asserted his moral right under the Copyright, Designs and Patents Act, 1988, to be identified as the author of this work.

www.gowerpublishing.com

British Library Cataloguing in Publication Data
Burtonshaw-Gunn, Simon A.
  Risk and financial management in construction
  1. Construction industry - Risk management 2. Construction
  industry - Finance
  I. Title
  624'.068

  ISBN: 978-0-566-08897-1

Library of Congress Cataloging-in-Publication Data
Burtonshaw-Gunn, Simon A.
  Risk and financial management in construction / by Simon A. Burtonshaw-Gunn.
    p. cm.
  Includes bibliographical references and index.
  ISBN 978-0-566-08897-1
  1. Construction industry--Risk management. 2. Construction industry--Finance. I. Title.
  HD9715.A2B873 2008
  624.068'1--dc22

                                                                            2008036687

Mixed Sources
Product group from well-managed
forests and other controlled sources
www.fsc.org Cert no. SA-COC-1565
© 1996 Forest Stewardship Council
FSC

Printed and bound in Great Britain by
MPG Books Ltd, Bodmin, Cornwall.

Contents

List of Figures   *ix*
Preface   *xiii*
Acknowledgements   *xv*

Introduction   1
    Book Format   1
    An Overview of the Construction Industry – Setting the
        Scene...   1
    Introducing Construction Risk Management   7
    Introduction to Finance and Investment of Construction
        Projects   10
    Book Overview: Part 1   11
    Book Overview: Part 2   14

PART 1   **Construction Risk Management**

Chapter 1 **Risk Management in Construction Projects**   21
    Risk Management and the Role of the Project Manager   21
    Overview of the Risk Management Process   24
    Risk Management in the Project Lifecycle   25
    Risk-Based Decision Making   28
    Risk Management and Culture   32

Chapter 2 **Risk Identification and Planning**   37
    Risk Classification   37
    Risk Identification   38
    Tools and Techniques for Risk Identification   42
    Risk Planning   45
    Appendix 1: Example Checklist for Risk Identification   56

Chapter 3   **Qualitative Risk Analysis and Quantitative Risk**
                 Evaluation   **59**
                 Qualitative Risk Analysis   59
                 Quantitative Risk Evaluation   65

Chapter 4   **Risk Response Planning, Monitoring and Control**   **73**
                 Risk Response Planning   73
                 Risk Monitoring and Control during the Project   79
                 Risk Monitoring and Control at Project Closure   83

Chapter 5   **Construction Prime Contracting and the Importance of**
                 Risk Management in International Projects   **89**
                 The Growth of Prime Contracting   89
                 Consideration and Identification of Risks on International
                    Projects   94

PART 2   **Financial Management**

Chapter 6   **Financing of Construction Projects**   **103**
                 Basic Economic and Financial Principles of Construction
                    Project Investments   103
                 General Criteria in Making Investment Decisions   108
                 Management of Risk and Uncertainty of Project
                    Investment   112
                 Methods of Raising Capital for Construction Projects   114
                 International Sources of Project Funding   117
                 Short- and Medium-Term Forms of Raising Capital
                    through Debt   118
                 Long-Term Forms of Raising Capital   120

Chapter 7   **Financial Assessment and Performance of Projects**   **123**
                 Financial Assessment of Project Investment   123
                 Financial Performance of Projects   125
                 Value Management Analysis   128
                 Tools and Techniques used in Value Management   130

Chapter 8   **Advances in Contract Strategy**   **137**
                 Introduction to Contract Strategy   137
                 Contract Strategy in Practice   146

Chapter 9 **Estimating, Budgeting and Cost Control** **151**
Estimating 151
Methods of Estimating 153
Budgeting and Cost Management 157
Cost Control 159
Tools and Techniques for Cost Control 161

References *165*
Further Reading 166
Useful Websites 167

Glossary *169*
Index *179*

List of Figures

Figure I.1   Competitive Rivalry   5
Figure I.2   Integrating risk management with other project
              management functions   8
Figure I.3   Benefits of risk management   10
Figure I.4   Balancing risk control   12

Figure 1.1   Typical corporate governance model   23
Figure 1.2   Balancing risk and control   23
Figure 1.3   Common steps in a risk management process   25
Figure 1.4   Correlation between construction process and risk
              management   26
Figure 1.5   Risk management process   29
Figure 1.6   Standard process for decision making   30
Figure 1.7   Risk response actions   31
Figure 1.8   Relationship of risk on strategic management   33
Figure 1.9   Culture and performance   34

Figure 2.1   Risk management tools   38
Figure 2.2   Typical project management risks   39–41
Figure 2.3   Tools and techniques for risk identification   42–44
Figure 2.4   Elements of risk management planning   45
Figure 2.5   A typical risk management plan   48–50
Figure 2.6   Project plan rationale   51
Figure 2.7   Probability and impact definitions   52
Figure 2.8   Probability/impact matrix   53
Figure 2.9   Table of impact definitions   53
Figure 2.10   Probability impact matrix (Risk score = P × I)   54

Figure 3.1   Qualitative risk analysis process   60
Figure 3.2   Risk severity analysis   61
Figure 3.3   Aggregating criticality scores   62

Figure 3.4   Example combined risk matrix   63
Figure 3.5   Ishikawa diagram   64
Figure 3.6   A typical FMEA table   65
Figure 3.7   Quantitative risk analysis process   66
Figure 3.8   Elements of a bow-tie diagram   67
Figure 3.9   Part of an FMECA matrix   68
Figure 3.10   Format of a risk register   69

Figure 4.1   Risk response actions   75
Figure 4.2   ALARP risk zones   78
Figure 4.3   ALARP principles   79
Figure 4.4   Elements of the risk monitoring and control process   80
Figure 4.5   Risk monitoring and control: the use of a closed loop
              feedback system   81
Figure 4.6   Responsibility matrix: post-project review process   87

Figure 5.1   Risk and supply chain management   91
Figure 5.2   Construction prime contracting and its relationship
              with other management disciplines   92
Figure 5.3   Areas of international risks   97

Figure 6.1   Phase 1 process for project investment   104
Figure 6.2   Phase 2 process for project investment   106
Figure 6.3   Phase 3 process for project investment   107
Figure 6.4   Percentage share of costs during the venture
              implementation cycle   110
Figure 6.5   Influence of various parties on the cost of new facilities   111
Figure 6.6   Sources of project funding   114

Figure 7.1   Triangle of project variables of time, cost and quality   123
Figure 7.2   Triangle of project variables of time, cost, quality and
              risk   125
Figure 7.3   Forward projections at different interest rates   126
Figure 7.4   Discounting back at different interest rates   126
Figure 7.5   Project 'S' curves showing cost and time relationship   132
Figure 7.6   Formulae used to calculate earned value   136

Figure 8.1   Customer-supplier risk allocation   142
Figure 8.2   Characteristics of different types of construction
              contracts   143
Figure 8.3   Risk insurance options   144

Figure 8.4   Contract cash flow projection – milestone management   147
Figure 8.5   Evolution to PPP   149

Figure 9.1   Contributors and problems of cost estimating   152
Figure 9.2   Variable and fixed costs   154
Figure 9.3   Example of three-point estimate for delivery of
                   materials to site   156
Figure 9.4   Cost profile   159
Figure 9.5   Cash flow   159
Figure 9.6   Project cash flow   160
Figure 9.7   Typical cost/time 'S' curve   162

Preface

The intention of writing this book has been to produce a reference tool for both practicing Project Managers and those studying the elements of Risk and Financial Management as part of their wider professional studies. Speaking from my own experience in a range of construction project sizes, locations and stages ranging from new build, refurbishment, modernization and demolition – high on the agenda for every Project Manager to consider is project risk and its associated costs. The related topic of project cost is important; not just throughout the project's duration but even before contract award, especially with an increase in self-funded infrastructure projects. Having said that, there are important lessons that can also be learnt at the project completion stage for both individuals and organizations; although so often the pressure to move quickly to the next project inhibits the value that such reflection and learning from experience can offer.

I have to confess that for me the proverb *'a picture is worth a thousand words'* applies to my learning style, finding the use of models, charts, tables and figures very valuable. I am sure that, because of the nature of project management in its use of Gantt charts, Work Breakdown Structures, programmes, risk matrices, reporting formats, spreadsheets and so on, I am not alone in this visual preference which is reflected in the following chapters. Whilst some of the chapter topics naturally lead one to another, I have tried as much as possible to provide a small amount of repetition allowing the reader to consider each topic area individually without having to have read the whole book in a mechanistic process way. Additionally I hope that the use of figures, tables and bullet point lists encourages the reader to dip in and out of the text to suit both their project requirements and also support and supplement their already acquired level of project management knowledge and experience.

Whether an experienced or a newly appointed Project Manager, or a student with ambitions in this discipline, I hope that you find this book of value in not just supporting your organization's projects but also in your own development.

Simon A. Burtonshaw-Gunn
Cheshire, UK.

Acknowledgements

I would like to record my thanks to a number of people who have helped bring this book into existence.

Firstly to my friends at the University of Salford for allowing me to be involved in the School of the Built Environment's modular MSc degree in Project Management; in particular Professor Chris Fortune, Mr Aled Williams, Dr Niraj Thurairajah and, of course, the Head of SoBE, Professor Mel Lees.

Secondly to those with industry experience who were able to offer constructive comments on the book as it developed, with particular thanks to Mr Alan Hoy, having high-jacked his free time to read the early Risk Management manuscript and to Project Management guru Mr Dennis Lock for his support, help and encouragement.

Thirdly, my thanks to publishers Ashgate, Gower, John Wiley and Sons, Blackwell Publishing, Free Press and PMI who allowed me to use previously published models to support this book. Where work has been previously published I have, where possible, obtained copyright permission for their reproduction, some figures, however, just exist in the public domain and as such, for these, it has not been possible to ascertain ownership. My thanks also to Mr Jonathan Norman and Ms Fiona Martin at Gower Publishing for their help and encouragement in progressing this project and to Mrs Charlotte Parkins for her superb editing skills.

Finally, as ever, to my wife Carole for her understanding of seeing me at the keyboard on evenings and weekends – sometimes going on well into the night.

Introduction

Book Format

Whilst there are natural overlaps between the two topic areas of Project Management, for ease of use this book is divided into two parts: Part 1 dealing with Construction Risk Management, and Part 2 with its corresponding Financial Management. This introduction provides an overview of the construction industry and also introduces the main themes of these two interrelated construction management topics. It closes with an overview of the forthcoming chapters.

An Overview of the Construction Industry – Setting the Scene...

The construction activity involves assembling materials and components designed and produced by a multitude of suppliers, working in a diversity of disciplines and technologies, in order to create what is regarded as 'the built environment'. Such activities can include the planning, regulation, design, manufacture, construction, maintenance and eventual decommissioning of buildings and other structures. Their scale, complexity and intricacy varies enormously, ranging from work undertaken by small 'jobbing' builders, to international construction companies undertaking long-term, high-cost, complex and sometimes high-risk projects such as single or multiple major civil construction, with obvious examples being the Channel Tunnel, nuclear power station construction and, of course for the UK, its showcase sporting venues for the London 2012 Olympic Games.

From a 'value chain' viewpoint, new construction can be regarded as:

- a means of production or provision of services (a factory or office block);

- an addition to, or improvement of, the infrastructure of the economy (railway or roads);

- a social investment (hospitals); or

- provision to meet a direct need (housing).

Construction then is a major national investment, accounting for half of the UK's annual fixed capital formation. As such, it plays a vital role in the competitive delivery of goods and services by the rest of the economy. In addition the house-building industry also stimulates other services (estate agents, solicitors) and associated industries for example suppliers of carpets, curtains, furniture, white goods and so on. The construction industry as a whole presents many employment opportunities in the fields of building, civil engineering, offshore structures and the process-plant industry. It embraces the efforts of contracting firms, specialist contractors, consulting architects and engineers, professionals such as architects and quantity surveyors, suppliers of building materials and manufacturers of equipment.

Construction, by its nature as a system integrator and a stimulus for other parts of the economy, can therefore be regarded as a basic economic multiplier. From a macroeconomic perspective the industry requires the three classic 'factors of inputs' of land, labour and capital, all of which can be affected by government policy. The UK is neither a state-controlled economy, nor is it at the other end of the continuum of being a free economy of no state intervention. Indeed the UK operates by a compromise of mixed economy with a certain amount of government intervention. At the national government level this can be seen in the form of taxes, Health and Safety legislation, Construction Management legislation and general employment law. At the local government level this is exemplified by control of planning approval and development schemes, adherence to local requirements and bye-laws and cognisance of local pressure groups.

Looking broader, the construction industry is not only of major national importance but may also feature as an international industry as UK construction companies may also be responsible for a significant amount of work undertaken overseas, which typically represents between 10 to 15 per cent of the annual turnover of the major British construction contractors. UK companies working overseas require an understanding of, firstly, how to find and undertake construction work in different countries; secondly, the level of competition, tendering and procurement activities; and thirdly, how to comply with new local and national government controls such as employment law, health and safety requirements and so on.

An examination of the various sectors within the construction industry reveals that this may be seen as a temporary activity as it is normally carried out at the client's premises and is viewed as an enabling activity to allow the client organization to conduct its business; be this retail, manufacturing or service related. However, because of these reasons, the industry has a number of problems such as low and discontinuous demand, low productivity compared to other industries and low profitability. There is no imposed or self-regulation of the industry and there are so many companies that the industry can be regarded as (and is often criticized for) being fragmented.

Historically the factors which influence the location of a particular industry have typically included the accessibility of raw materials, power supply (for example, coal and water), transport, market access and so on. However, examining the construction industry against such criteria reveals that the industry is not itself influenced by such location factors but by the locations of its client's base and, without a long-term commitment, construction work becomes a mobile service industry. Indeed, in the vast majority of cases it is not only undertaken at the client's site but, importantly, only when other industrial or commercial sectors are flourishing, allowing it to concentrate on the building of factories, houses or service buildings.

The operation of the construction industry can be viewed against a competitive strategy model (shown as Figure I.1) developed by Harvard University Professor, Dr Michael Porter, which identified five key driving forces on a business. These being:

- Potential Entrants – the threat of new entrants;

- Industry Competitors – rivalry among existing firms;

- Substitutes – the threat of substitute products or services;

- Buyers – the bargaining power of buyers;

- Suppliers – the bargaining power of suppliers.

In looking at the first two of these key business drivers – Potential Entrants and Industry Competitors – it is recognized that the UK construction industry is, in the main, confined to competition from established UK companies. This is particularly so in the sectors of Non-residential Building and Infrastructure because the construction industry operates differently to other industrial or economic sectors. Companies require manpower resources, plant and equipment, health and safety procedures and various insurances to be in place together with an established track record to be able to even bid for work in this important sector. Consequently, these requirements imposes significant barriers to entry to the construction market resulting in competition not from international companies or new entrants but being confined largely to existing UK companies or those international organizations with a UK subsidiary operation. In looking at the third element – the threat of substitute products or services – it is logical to assume that there is little threat of substitution of site-based infrastructure construction activities although the future may see an increase in the prefabrication of buildings or other factory-based building techniques.

The final two forces covering the bargaining power of buyers and those of suppliers can also be grouped together. Both are influenced by the same factors of the concentration of construction companies, the number of capital projects being undertaken within the UK market and the perceived importance and willingness to undertake work for a particular client. These two forces and that of competition have had an effect on client/contractor relationships over a period of many years. Indeed, the traditional *modus operandi* of competition with other construction companies has been driven, in the main, by the client's desire to achieve the lowest-cost bid, not only in the private sector but also in the public sector, which has been supported by the government-inspired Compulsory Competitive Tendering process. As a result, and with very few exceptions, relationships have been adversarial with parties concerned resorting to contractual claims, which lengthen timescales and drive up costs, as a direct consequence of the behaviour of both parties. Where the client will only award work to the lowest-priced contractor, the contractor responds to this approach by submitting an uneconomic and unrealistic low price which results in time and effort being spent by the contractor in finding reasons for which additional money will be sought. In essence, this became a common contractor's traditional strategy of submitting a low price in competition knowing that it will be supplemented by post-contract extras, variations and financial claims when delay and disruption opportunities presented themselves.

Furthermore the traditional procurement systems and the contractual and legal framework by which participants are bound together have been widely and regularly criticized as being confrontational and adversarial. This should not be too surprising as it is widely recognized that the construction industry is historically famed for its adversarial approach and the amount of time and

Entry barriers

- Economies of scale
- Proprietary product differences
- Brand identity
- Switching costs
- Capital requirements
- Access to distribution
- Absolute cost advantages-
    - proprietary learning curve
    - access to necessary inputs
    - proprietary low-cost product design
- Expected retaliation

Rivalry determinants

- Industry growth
- Fixed (or storage) costs/value added
- Intermittent overcapacity
- Product differences
- Brand identity
- Switching costs
- Concentration and balance
- Informational complexity
- Diversity of competitors
- Corporate stakes
- Exit barriers

New entrants

Threat of entrants

Suppliers — Bargaining power → **COMPETITIVE RIVALRY** ← Bargaining power — Buyers

Determinants of suppliers

- Differentiation of inputs
- Switching costs of suppliers and firms in the industry
- Presence of substitute inputs
- Supplier concentration
- Importance of volume to supplier
- Cost relative to total purchases in the industry
- Impact of inputs on cost or differentiation
- Threat of forward integration relative to threat of backward integration by firms in the industry

Threat of substitutes

Substitutes

Determinants of substitute threat

- Relative price performance of substitutes
- Switching costs
- Buyer propensity to substitute

Determinants of buyer power

- Bargaining leverage
- Buyer concentration v firm concentration
- Buyer volume
- Buyer switching costs relative to firm switching costs
- Buyer information
- Ability to backward integrate
- Substitute products
- Pull-through
- Price sensitivity
- Price/total purchases
- Product differences
- Brand identity
- Impact on quality/performance
- Buyer profits
- Decision makers incentives

Figure I.1   Competitive Rivalry

Source: Reprinted with the permission of The Free Press, a Division of Simon & Schuster Adult Publishing Group, from '*Competitive Advantage: Creating and Sustaining Superior Performance*' by Professor Michael E Porter. © 1985, 1998 Michael E Porter. All rights reserved.

money spent on litigation which provides employment opportunities for a significant number of professionals including claims consultants, quantity surveyors and construction lawyers. Such recognition of the construction industry's adversarial attitudes is not new and a number of British Governments have commissioned investigations into the industry, notably the reports of Sir Ernest Simon in 1944, Sir Harold Emmerson in 1962 and Sir Harold Banwell in 1964, all of which tell a similar tale. More recently this industry 'norm' so concerned the previous Government that it appointed Sir Michael Latham to carry out yet another review of the procurement and contractual arrangements in the industry. His report '*Constructing the Team*', published in

1994, recommended change for the industry and gave a number of specific recommendations including a change in culture aiming to move away from the traditional adversarial client/supplier relationship which it noted had prevailed in the industry for many years. In particular the *'Constructing the Team'* report saw closer collaboration and 'partnering' as the development of a strategic long-term commercial arrangement between a client, contractors and suppliers and went further by proposing that this was one of the techniques most likely to improve both cost efficiency and customer relationships within the construction industry. However, the traditional industry relationship between client and contractor over a period of many years had led to a set of assumptions that had become strong barriers to implementing partnering relationships in this industry.

These adversarial attitudes resulting from traditional contractor/client relationships often concern the priorities and goals of the client which in the past have been substantially different from those of the contractor. This was also due to the involvement of the client who appointed a number of consultants on a construction project to design and supervise its construction. Often there would be several such consultants on a single project, including architects, civil and structural engineers, mechanical, electrical and services engineers and quantity surveyors with one of these consultants appointed to undertake the role of managing and coordinating the work of the others and the contract as a whole. Because of the many specialist groups often involved in a project, clients have expressed concern that they have limited communication of progress and little, or no, control over the costs of construction, nor of the time taken to complete the project. Such comments arose from the past reluctance to involve the client in the decision making process with a preference to keep any technical problems concealed from the client until the final payment demand was presented. In addition to poor communications clients had also felt that they are the victims of poor design, inadequate supervision, insufficient choice of materials and contract methods. To counter such problems major clients, including the UK Government itself, demanded a change in the construction industry's behaviour with a change in emphasis from a 'production-oriented' to a 'client-oriented' outlook. This viewed was echoed in the major industry reviews conducted by Sir Michael Latham and Sir John Egan in 1994 and 1998 respectively where such culture change was fundamental for work being undertaken in a partnering arrangement promoted by the Latham and Egan reviews. Over the last decade some government agencies, and other clients, have had encouraging experiences in partnering, although the government still does not define the term 'partnership' in quite the same way as the construction industry. For them partnership means, almost entirely, getting the private sector to employ its resources and assets to take on as much risk as possible including the risk of cancellation or postponement, often for little financial reward.

Against this industry backdrop, construction projects require the deployment of project managers and their professional skills and experience to bring the project to fruition. This role involves managing the interrelated project performance parameters of cost, time and quality and managing the risks which may adversely affect one, two or all three of them.

Introducing Construction Risk Management

In life there are risks: in driving a car, crossing the road or playing various sports. Everything we do can be associated with risk in the form of events that might prevent the achievement of stated objectives if they occur. So too in business although in many cases such risk uncertainties are naturally associated with a financial risk compared to the market volatility and hence the ability to realistically provide expectations based upon a risk versus reward trade off. Whilst the management of corporate financial risk is undertaken through a very specialist risk discipline, this book examines the subject of risk management from a project, business or operational viewpoint where such risks can be internally or externally driven and may impact on the project's stated scope, schedule and cost objectives.

The objectives of risk management are to ensure the rapid identification of risks within the business and to establish a clear process of assessment, action planning and reporting of the risks identified. In addition, it is important that focus and attention is given to the identification of opportunities as this will enable effective decision making to ensure that:

- Business opportunities can be quickly assessed at an appropriate level in order to decide whether and how it might proceed with such opportunities.

- Threats to the project or other parts of the company's operations can be eliminated or at least reduced to an acceptable level.

- All decisions take account of contributing to sustainable shareholder value.

The underlying principle is that key risks and the appropriate control measures are kept under regular review and reported to project participants, project sponsors and key client representatives.

In focusing on typical construction projects, the topic of risk management can be seen in Figure I.2 to impact on many facets of the project. Whilst the traditional view is that risk management is a part of the project management function, carried out by the project manager or delegated project team member, an alternative view is that if there were no risks in a project there would be no need for project management and that the main purpose of project management is to manage the risks and hence the term Risk-Driven Project Management has started to come into being. From this understanding, risk management should consider all aspects of the project and begin early in the life of a construction project continuing through until project closure.

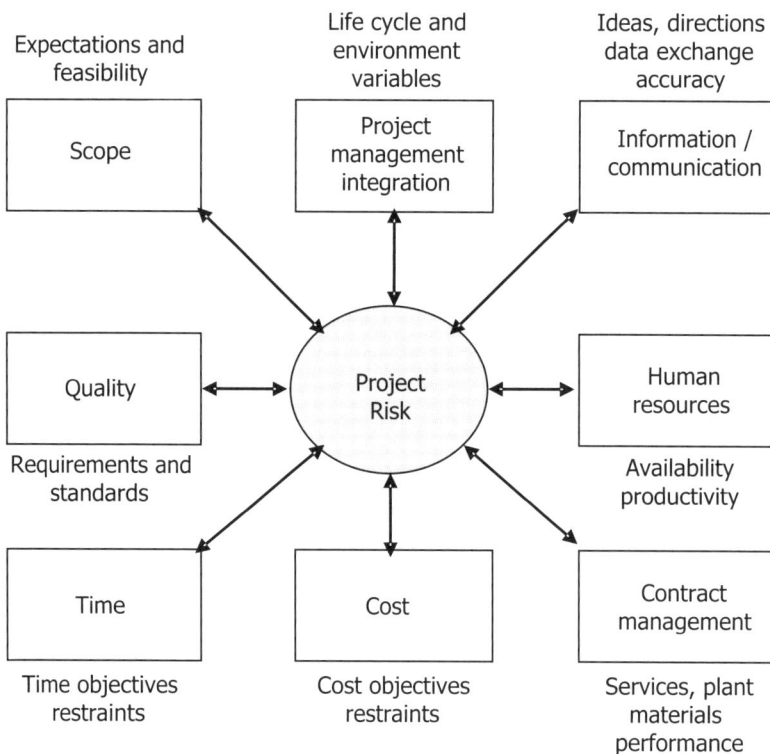

Expectations and feasibility	Life cycle and environment variables	Ideas, directions data exchange accuracy
Scope	Project management integration	Information / communication
Quality	Project Risk	Human resources
Requirements and standards		Availability productivity
Time	Cost	Contract management
Time objectives restraints	Cost objectives restraints	Services, plant materials performance

Figure I.2   Integrating risk management with other project management functions

A project by definition is trying to introduce some form of change – a new production system or way of working, a new building, and so on. Change involves uncertainty, which in turn means that projects are more likely to be 'blown off course' by a potential future event. In other words, projects in themselves are inherently risky undertakings. Dennis Lock, in introducing the topic of risk management in his book, *Project Management*, proposes that:

> *'It is not surprising that projects, which metaphorically (and sometimes literally) break new ground, attract project risk. Project risks can be predictable or completely unforeseeable. They might be caused by the physical elements or they could be political, economic, commercial, technical or operational in origin. Freak events have been known to disrupt projects, such as the unexpected discovery of important archaeological remains or the decision by a few members of a rare protected species to establish their family home on what should have been the site of the project.'*

There is widespread agreement within the project management community that a project risk is any event or series of events, whether internally or externally driven, that on occurring will have negative consequences on the project or business opportunity in terms of performance, functionality, time of delivery, acceptance or cost. There are, however, strong views that project risks are always those risks that impact on one or more of the project baseline elements – time, cost or quality (note that quality is sometimes referred to as technical).

The British Standard on Project Management (EN BS 6079-3:2000) defines risk as:

> *'An uncertainty inherent in plans and the possibility of something happening (i.e. a contingency) that can affect the prospect of achieving business or project goals.'*

In the Vocabulary of this Standard a further definition is also provided as a:

> *'combination of the probability or frequency of occurrence of a defined threat or opportunity and the magnitude of the consequences of the occurrence.'*

Another common definition of risk and one frequently used is:

> *'the threat or possibility that an action or event will adversely or beneficially affect an organization's ability to achieve its objectives.'*

There are many potential benefits to the effective use of risk management techniques, the most significant are shown in Figure I.3 below.

Introduction to Finance and Investment of Construction Projects

Whilst the first part of the book deals with project delivery risks there are also many risks associated with financing of large-scale investments, both by private and public organizations. Every project requires financial means, regardless of whether it is a public, public-private or a privately-funded venture and investors are often afraid of making decisions due to lack of full knowledge in the field of financing methods and their associated risks. As such, the second part of this book will deal with the methods of financing, assessment of effectiveness and identification and management of risks while undertaking construction investment projects.

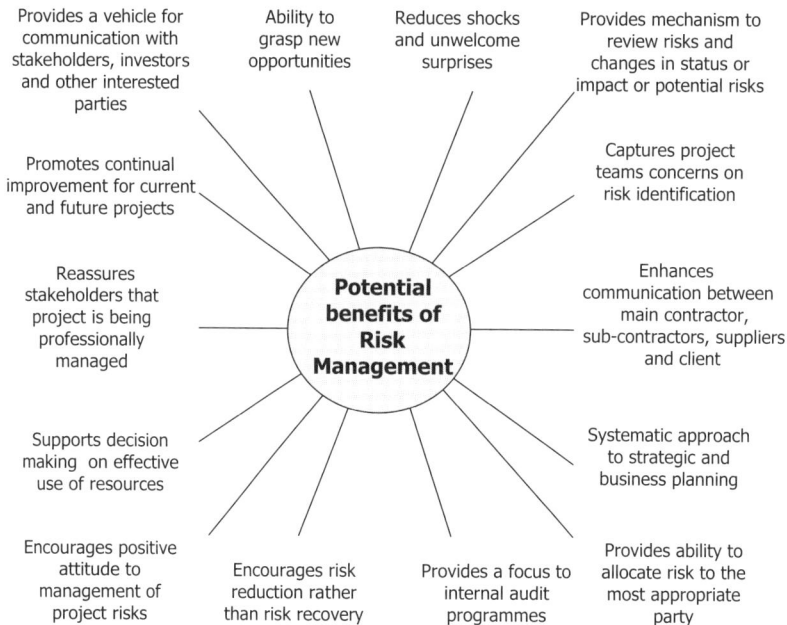

Figure I.3   Benefits of risk management

All organizations will expect the potential investor to consider risks inherent in large-scale ventures, both those risks associated with the construction and operation and those pertaining to the mode of financing of the project. Many projects undertaken in developing countries are financed by multilateral, bilateral or special purpose financial organizations. In most cases, however, a lack of developed financial markets often limits financing to only a few limited sources. In countries characterized by a highly developed market economy, both the state and private companies can be seen to finance numerous undertakings. Here the sources of private finance have traditionally come from pension funds, insurance companies, commercial banks, niche investment banks, large corporations, stock exchanges, assistance agencies, investors and property (real estate) developers.

The sources of finance from the public sectors almost wholly arise from taxation or subventions to provide the required level of financing of the project. In regression periods only urgent projects may be financed this way since income from taxation is likely to be limited and subventions, often in form of interest-free loans, are then the only option.

Modern economic analysis and finance management methods provide numerous solutions that allow investors to make the right decisions throughout all phases and stages of the investment process. Specific groups of methods also allow for management of uncertainty and risk while making investment decisions. In the case of construction projects, the main problems are in the estimates of their profitability and ensuring the proper modes of financing. After securing the project from a client where funded by internal, external or the client, project cost control and financial performance will be key areas to be managed.

Book Overview: Part 1

This book consists of two parts; in the first part are five chapters on risk management which are presented in a natural sequence of the components of the topic of risk management. These chapters are introduced below:

CHAPTER 1: RISK MANAGEMENT IN CONSTRUCTION PROJECTS

In introducing the topic of risk management it should be noted that whilst there will be a cost to the project in adopting a risk management process and indeed culture, this cost can also be considered with respect to the consequences of *not* undertaking risk management in a systematic and professional way. Clearly all risks cannot be controlled but to ignore risks and risk mitigation tools, covered in this and the subsequent chapters, will undoubtedly lead to adverse consequences on projects. Such consequences of failing to deal effectively with risk can include significant cost overruns, schedule delays and inability to achieve desired project technical objective(s). Other important consequences may include:

- project de-scoping;
- loss of credibility;
- ultimately, project cancellation and unhappy clients; and
- personal or organizational liability and fines.

The balance between the organization's ability to take risks for business purposes and that of risk management in the form of corporate governance and a management process is shown in Figure I.4 which illustrates the difficulty of balancing risk taking with risk management; in reality these two forces inevitably cause the project to osculate around the organization's optimum approach.

Chapter 1 looks at the subject of risk management from a project, business or operational viewpoint. Its aim is to provide an introduction to the topic where such risks can be internally or externally driven and have the ability to impact on a project's stated scope, schedule and cost objectives. The role of risk management and that of the project manager are discussed in this chapter,

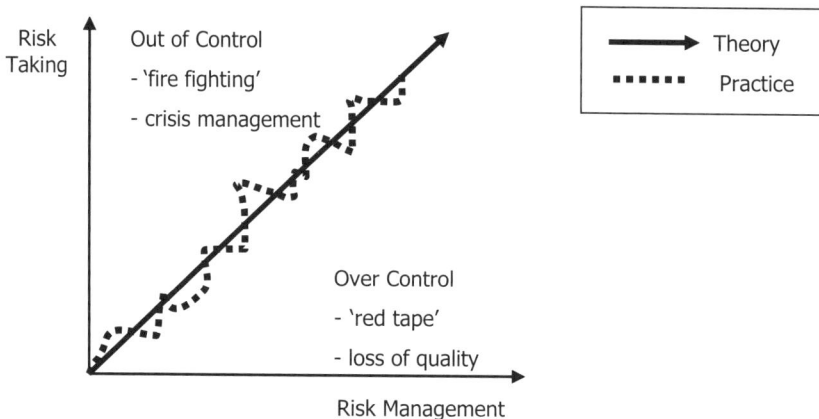

Figure I.4   Balancing risk control

Source: From '*The Essential Management Toolbox*', S.A. Burtonshaw-Gunn (2008). © John Wiley and Sons. Reproduced with permission.

followed by an overview of the risk management process and where it is used in the project process.

This chapter also provides a risk identification checklist as an appendix.

CHAPTER 2: RISK IDENTIFICATION AND PLANNING

This chapter covers the important task of risk identification which is traditionally considered to be the first step in the risk management process and includes tools and techniques to aid this process. Once the risks have been identified then a Risk Plan can be developed and Chapter 2 discusses both the rationale and suggested contents of a typical risk plan. The chapter concludes with a look at the definitions of risk probabilities and their consequences which are key features in efficiently using resources to implement the risk plan.

CHAPTER 3: QUALITATIVE RISK ANALYSIS AND QUANTITATIVE RISK EVALUATION

Chapter 3 covers two main topics. Qualitative risk analysis covers a range of techniques for assessing the impact and likelihood of identified risks. These approaches can be used to prioritize the risks according to their potential effect on project objectives and is one way to determine the importance of addressing specific risks and guiding risk responses. The time-criticality of risk-related actions may magnify the importance of a risk and together with an evaluation of the available information also helps modify the assessment of the risk. Quantitative risk evaluation generally follows the qualitative risk analysis activity. The second part of the chapter shows that quantitative risk evaluation requires risk identification and that both qualitative and quantitative risk analysis processes may be used separately or together. There are many tools available for the identification and evaluation of risks and risk controls, ranging from experience-based judgement, checklists and risk matrices, to specialist review and analysis techniques. The most appropriate tool depends on the operation complexity and level of risk; and the ease of use and form of output. This chapter also covers the more specialist risk tool 'Bowtie' methodology which is used in a number of the high-risk industries and is a very popular tool in the oil and gas sector.

CHAPTER 4: RISK RESPONSE PLANNING, MONITORING AND CONTROL

Risk response planning is the process of developing options and determining actions to enhance opportunities and reduce threats to the project's objectives.

This includes the identification and assignment of individuals and parties to take responsibility for each agreed risk response. This process ensures that identified risks are correctly addressed and that the effectiveness of response planning will directly determine whether the risk increases or decreases for the project. There are a number of factors that risk response planning must satisfy, namely:

- its appropriateness to the severity of the risk;

- that it needs to be cost effective in meeting the challenge;

- that it needs to be timely to be successful;

- that it needs to be realistic within the project context;

- that it needs to be agreed upon by all parties involved in the project; and,

- that it needs to be owned by a responsible person.

Selecting the best risk response from several options is often required and these are discussed in Chapter 4 together with risk monitoring and control in order to keep track of the identified risks, monitoring residual risks and identifying any new risks as they emerge during the life of the project. The chapter covers a typical project control system to ensure the execution of the risk plan and evaluating the plan's effectiveness in reducing risk. Risk response planning also needs to be monitored against the risk plan and may involve choosing alternative strategies, implementing a contingency plan, taking corrective action(s), or, at worst, replanning the project.

CHAPTER 5: CONSTRUCTION PRIME CONTRACTING AND THE IMPORTANCE OF RISK MANAGEMENT IN INTERNATIONAL PROJECTS

Having examined risk management for all construction projects, this final chapter on risk management commences by considering risk management applied to large-scale projects. It also introduces the topic of Prime Contracting and goes on to conclude with an examination of the risks likely to be prevalent in international construction projects.

Book Overview: Part 2

The second part of the book covers four chapters on financial management commencing at the strategic financial level of raising project funding, through to project performance contract strategy and on to project financial aspects of estimating, budgeting and cost control. These financial chapters are introduced below:

CHAPTER 6: FINANCING OF CONSTRUCTION PROJECTS

The first chapter of the second part of the book now represents a move to financial management with a good starting point being to look at how investment in construction projects is undertaken at various stages of the project. The chapter then goes on to discuss the influences on, and the criteria for, making such investments together with some reference to risk and uncertainty of project investment.

Whilst some construction clients may look to external investors to provide funds for the construction of their business facilities, they are also likely to use their own funds in addition to such externally raised capital. Where the client company does not have enough immediate funds from either its normal business income activities, or even by raiding its 'piggy-bank' of retained profits it is unable to provide sufficient funding for large-scale investment, the options to raise and/or increase initial project funding for construction projects in both the long and short term are explored in this chapter.

CHAPTER 7: FINANCIAL ASSESSMENT AND PERFORMANCE OF PROJECTS

Following on from Chapter 6 and the project performance criteria of cost, time and quality is the financial assessment of the project investment and a number of techniques for examining the financial performance of projects including Value Management.

CHAPTER 8: ADVANCES IN CONTRACT STRATEGY

This chapter examines construction contract strategy with a focus on risk and insurance providing a link back to the earlier chapters of the book. This chapter ends with a look at the financial aspects of Public-Private Partnerships – a contract strategy growing in popularity especially for large-scale infrastructure projects.

CHAPTER 9: ESTIMATING, BUDGETING AND COST CONTROL

The final chapter covers the main financial functions of project delivery. Project management requires accurate cost estimates; if the project information used is detailed and precise, the resulting cost estimate will be also. Indeed during the duration of the project, as the project technical scope of work becomes defined so the precision of the estimate will improve. Although individual companies will use their own estimating processes usually these will follow a typical approach, such as:

- Obtaining specific technical and operational concept definition to understand what the project's objectives are and how will they be accomplished.

- Developing the Work Breakdown Structure (WBS) and input on how the project be organized.

- Building a structured cost chart of accounts to allow the costs to be categorized.

- Building a data collation plan and identifying what data is needed, and how it will be collected.

- Data collection: obtaining and recording the required information.

- Structuring the cost database with consideration on how the data is to be organized for best developing cost estimates.

- Data analysis to understand what the data indicates about the project costs.

- Developing a cost estimating relationship to determine if certain costs can be forecast as a function of the project's parameters.

- Building an interactive cost model to represent the project costs which will demonstrate how overall project costs change as various conditions change.

- Making an estimate of the costs based on expectations and project conditions.

- Analyzing the results to ensure that the cost estimate reflects any changes in the project conditions.

The chapter details these necessary cost control functions and provides guidance on the three main areas of the chapter title.

REFERENCES

Where material has been taken from other publications this is detailed in the references section to assist readers if further detail is required.

GLOSSARY

It is considered useful to include a glossary covering terms used in the topics from both parts of this book.

PART 1
Construction Risk Management

Risk Management in Construction Projects

CHAPTER

1

Risk Management and the Role of the Project Manager

[handwritten: Risk is both positive & negative.]

Risk Management is a means of dealing with uncertainty – identifying sources of uncertainty and the risks associated with them, and then managing those risks such that negative outcomes are minimized (or avoided altogether), and any positive outcomes are capitalized upon. The need to manage uncertainty is inherent in most projects which require formal project management. In looking at risk management and the role of the project manager it should be noted that risk management cannot be owned by one individual on a project and that all team members must be 'risk aware' and participate in activities to improve a project's position, through Action Plans, which are part of the main Project Plan. The two objectives for the deployment of the discipline of risk management are:

[handwritten: Prevent risks from occuring.]

1. To plan and take management action to achieve the aims of removing or reducing the likelihood and effects of risks before they occur and dealing with actual problems when they do; and

2. To continuously monitor potential impacts of risks, review the associated action plans, and provide and manage adequate financial and schedule contingencies for risks should they occur. *[handwritten: React to risks.]*

To be fully effective, project managers need to recognize that risks exist and actively manage them; this should be viewed as an indication of good project management; not an admission of failure. By looking ahead at the potential events that may impact the project and putting actions in place to address them (where appropriate), project teams can proactively manage risks and increase the chances of successfully delivering the project within the time, cost and quality project requirements. Whilst, in the early days of project management, great emphasis was placed on managing cost and schedule adherence, in the 1980s companies recognized the need to integrate technical risk with cost,

[handwritten: risk =+]

schedule and quality risks and thus risk management systems were developed to become a key project management discipline.

It should be noted that the project manager's responsibility is not to make risks 'disappear' but to manage and communicate these through the implementation of a systematic risk management process. Often it is not possible for the project team to identify all risks as the unexpected may still occur, however these instances should be very rare and project staff should be familiar with dealing with other examples of risk occurrence and mitigation. All project staff will have some level of responsibility for internal control as part of their accountability for achieving both the project's and their own objectives. Indeed this is even true of those above the direct project teams, as The Institute of Chartered Accountants report:

> 'They, collectively, should have the necessary knowledge, skills, information and authority to establish, operate and monitor the system of internal control. This will require an understanding of the company, its objectives, the industries and markets in which it operates and the risks that it faces.'

> > Internal Control, Guidance for Directors on the Combined Code on the Committee on Corporate Governance, The Institute of Chartered Accountants

Looking at risk management from the most senior level of the organization, from December 2000 all London Stock Exchange listed companies have been required to comply fully with the Turnbull Report on Corporate Governance which notes that:

> 'Risk management is essential for reducing the probability that corporate objectives are jeopardised by unforeseen events. All that the company is trying to achieve can be affected by risk exposure. They should be proactively managed.'

> > Implementing Turnbull, A Boardroom Briefing The Institute of Chartered Accountants

The Institute of Chartered Accountants also reported on the above that:

> 'non compliance with the Turnbull guidance would result in an embarrassing disclosure in the annual report that could attract attention of the press, shareholder activists and institutional investors.'

Given the importance of the above issues on corporate governance and its relationship to being able to demonstrate responsible risk management, many of the larger organizations in the UK embed risk management within their corporate governance model with a typical framework approach being shown in Figure 1.1. This framework of corporate governance illustrates the philosophy that effective controls start with the company's executive in providing strong leadership with clear accountability. The purpose of the risk framework sets out the central philosophies for controlling the organization's activities. These are then traditionally underpinned by requirements regarding behaviours, ethics and compliance with the organization's policies.

Another view of relating risk management to the role or culture of the company is illustrated in Figure 1.2, which shows a clear relationship between an organization's objectives, risk and controls and its risk exposure. By proactively addressing risks correctly the project should cost less, be completed more quickly and produce products more likely to meet the client's requirements.

Senior Management Accountabilities		
Control Framework		**Risk Framework**
Behaviours	General and Business Ethics	Policies

Figure 1.1   Typical corporate governance model

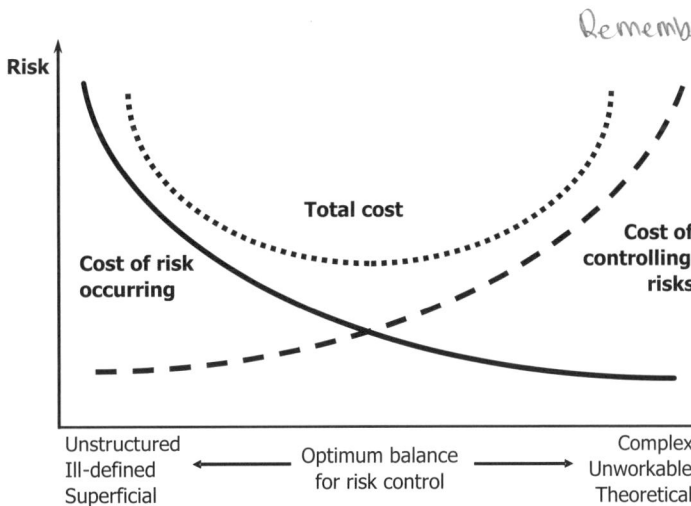

Figure 1.2   Balancing risk and control

As such the test of an organization's commitment to achieving effective risk management is the visible willingness to allocate budget or other resources to risk actions at each stage of the project. Although this may imply that all the actions are on the senior management of the organization, the responsibilities for risk management are far wider than this as both the project manager and project teams have direct project and governance responsibilities in adhering to the corporate risk requirements.

All organizations have a mixture of expressed or implied objectives. Risk management as a process for avoiding risk will actively support the achievement of those objectives, indeed when used well it can actively allow an organization to take on activities that have a higher level of risk (and therefore the opportunity to deliver a greater benefit), because the risks have been identified, are understood and are being actively managed. Risks can be managed through the operation of controls. However, even high levels of investment in control processes will not always eliminate all risk and any remaining risk becomes the organization's 'exposure' to risk. This is also known as its net or residual risk.

Overview of the Risk Management Process

Individuals and organizations need to have an all-round approach to risk management; the most progressive risk management organizations achieve this by a uniform balance of tools, process, attitude and SQEP (suitability qualified, experienced/knowledgeable personnel). Risk management has evolved into a formal systematic process of identifying potential risk or uncertainties and developing, selecting and managing options for addressing the risks throughout the life of the project. Risk evaluation is a key element of the corporate system for managing business risks with its main focus on identifying risks, evaluating their severity and managing the process.

Whilst risk management may be a proactive approach it cannot control future events but will allow decisions to be made and actions taken if such identified risks become reality. An understanding of risk comes from a realistic understanding of what can go wrong, the likelihood of the event occurring and the consequences of such an occurrence. On the basis of such an understanding a number of actions are open to organizations to manage the risk and, to a large extent, such response will be a function of the probability and quantified consequences of the risk occurring.

To support this process there are a number of common descriptions of the risk management process steps but in general they all follow a similar basic approach of risk identification, risk quantification, risk response and risk control, as shown in Figure 1.3 which has a similar four step approach of risk identification, assessment, control and recovery.

Clearly all risks cannot be controlled but to ignore risks and the risk mitigation tools covered in the following chapters will undoubtedly lead to adverse consequences on a project. Such consequences of failing to deal effectively with risk can include loss of credibility and may include personal or organizational liability and fines. Other important consequences include significant cost overruns, inability to achieve desired the project technical objective(s), schedule delays, project de-scoping and ultimately project cancellation. All of which are likely to lead to unhappy clients and a significant reduction in future project opportunities with the same client.

Risk Management in the Project Lifecycle

Different risks occur at different times or stages in the project and no investment or even short-term production process can be planned without taking into account the associated risk during its lifecycle. In reality every project contains a component of risk which results in the necessity to assess and reduce the associated threats. On this premise, risk management is a continuous process which should be conducted at every stage of the project; from its emergence through to completion and operational use. It is important to eliminate risks as early as possible; for instance at the stage of analysis of the project value and at the cost analysis stage during the project implementation. The main objective is to identify the problem as well as its significance; together with any associated benefits with the risk management process. This identification can be documented as part of the risk plan.

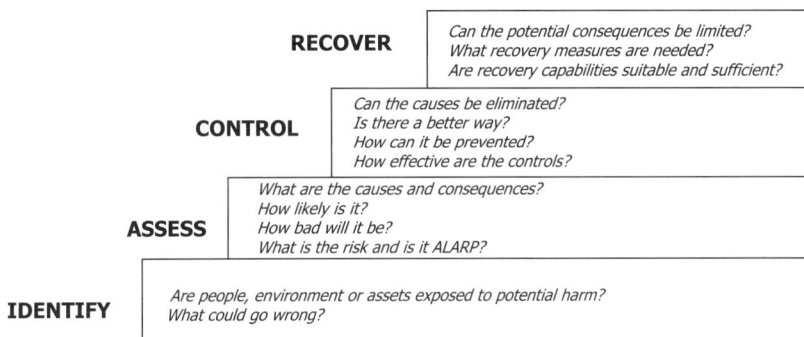

RECOVER
Can the potential consequences be limited?
What recovery measures are needed?
Are recovery capabilities suitable and sufficient?

CONTROL
Can the causes be eliminated?
Is there a better way?
How can it be prevented?
How effective are the controls?

ASSESS
What are the causes and consequences?
How likely is it?
How bad will it be?
What is the risk and is it ALARP?

IDENTIFY
Are people, environment or assets exposed to potential harm?
What could go wrong?

Figure 1.3   Common steps in a risk management process

Figure 1.4 presents a correlation between the various levels of risk management and the subsequent stages of the construction process. Although there are no rigorous divisions between the subsequent stages of the risk management process these are shown as the risk at company, strategic and project levels. It is worth remembering however that along with the progress of work, the approach to the problem of risk should remain under constant review which results in the necessity to consult with various specialists depending on the specific characteristics of the changing risk issues.

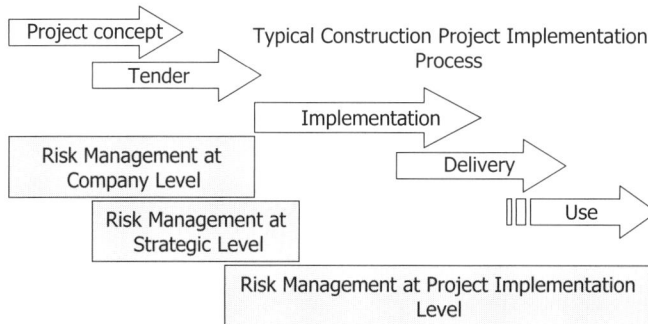

Figure 1.4   Correlation between construction process and risk management

With respect to construction projects it is suggested that risk management can be undertaken at a number of levels or stages of the project as shown in Figure 1.4 and described below.

Risk management at the Company Management level is the first stage in the risk management process where there should be a concentration on the elimination of threats in the general approach towards a given project investment. The basic elements, which should be taken into account at this stage of risk management, will typically include:

- perception of the problem by the client;

- requirements of clients;

- influence of taxes;

- repayment period;

- comparison of various approaches to investment;

- public procurement;

- initial assessment and identification of risks;

- depreciation;

- comparison of initial risk assessment with expected rate of return of capital;

- the expected cash flow;

- portfolio analysis;

- market trends.

The risk areas listed above will provide the company with an initial view on the project's attractiveness and prompt questions regarding whether the undertaking aligns to its business plan and shareholder's expectations.

Moving on to the second stage covering risk management at the Strategic level, this now needs to consider more specific problems such as the:

- contractor selection procedure;

- designer selection procedure;

- cost control methods;

- project management information systems;

- control of key issues of the project;

- control of critical path;

- selection of insurance;

- financing of the project;

- priorities of the project;

- result on the initial analysis.

Finally, risk management needs to be addressed at the Project Implementation level. This stage is associated with the work on the project and as such is associated to a greater extent with the participants of the design and implementation process. This final stage is one where the maximum risk management activities usually feature although some of these may remain as a legacy of the two previous stages. For many organizations this third stage encompasses their whole process of risk management which consists of more practical problems, such as:

- shortage of materials;

- accidents;

- weather;

- changes in orders;

- delays in designing;

- liquidation or bankruptcy of a contractor;

- additional cost of contractors;

- safety;

- quality of employees;

- unknown parameters of the construction site;

- safety and environmental protection;

- interpretation of the project.

This third stage, spanning project implementation, is for many companies the starting point of their risk management process commencing on award of the contract. It is suggested that the pre-contract risk management stages are more important especially for those construction projects undertaken as a Prime Contract away from the contractor's home country. This point is discussed in detail in Chapter 5.

In looking at risk management within the life cycle of a construction project, this topic needs to use robust plans that will survive in the practical sense and cover both known and unknown risks with respect to the project undertaking. Whilst the response to known risks are typically proactive, managed, planned and budgeted for; on the other hand, the response to unknown risks are often reactive, unmanaged, not budgeted or resourced and often encourage further unplanned reactions.

Risk-Based Decision Making

Many organizations in commerce, industry and the public sector have learnt the need for structured Risk-Based Decision Making processes after some very painful lessons. Few of these would state their processes are fully evolved and functioning without problems. Many other organizations are really only now starting a similar journey.

Risk-based decision making, which itself provides a framework for the typical risk management process approach, is shown in Figure 1.5, of which variations of this are in universal use for construction risk management. Successfully applied though, risk-based decision making can be both powerful and cost effective.

The steps in the risk management process shown in Figure 1.5 start with identifying the potential risk and recording this in a formal way, this is described in detail in the next chapter. This is followed by the assessment of the risk, looking to qualify and quantify the risk against a set of consistent criteria using various analysis techniques to determine the risk significance and the means for managing its consequences. This part of the process is described in more detail in Chapter 3. Decisions evolve around the need to make choices, either to do or not to do something, or to select one option from a range of possibilities. The choices available are often constrained, or at least influenced by, a range of social, technical, business, safety and environmental requirements and objectives. Successful decision making not only requires an understanding of these many requirements and objectives, but also an understanding of their relative importance and how to assess options and make the 'best' decision. A standard framework for the decision making process is illustrated in Figure 1.6 where the importance of the change dictates the extent and formality of the assessment, documentation, review, consultation and approval stages.

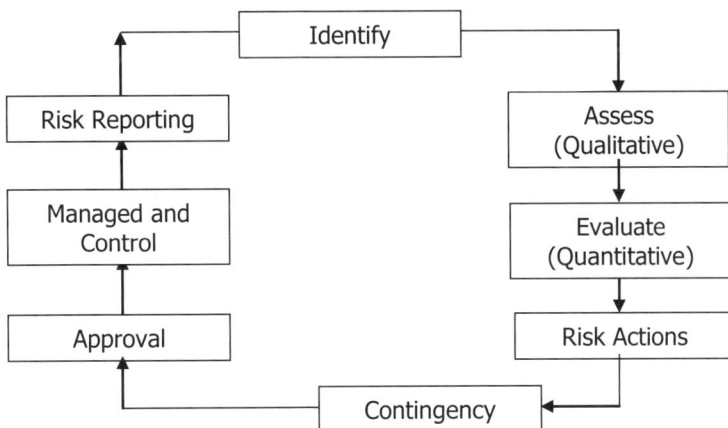

Figure 1.5   Risk management process

With the model shown in Figure 1.6, the need to change could come from a number of sources, including statutory requirements, internal reviews, audit findings or lessons learned form major incidents, for example. The extent of assessment and documentation will be dependent on the significance of the proposed change and is likely to range from assessment based on past experience, knowledge of similar projects or knowledge of the client requirements and its culture, through to more comprehensive numerical assessment depending on the complexity of the project and the cost, time and quality implication of the risk occurring. Typically, the extent of review will be dependent on the significance of the proposed change and as such may be undertaken from internal reviews, independent external reviews and finally (but only in extreme circumstances) the involvement of regulatory bodies, such as the UK's Health and Safety Executive. Implementing the change effectively is arguably the most important step, since it is only at this point in the process that the action is able to reduce the identified risk.

Following on from the standard decision making process, the overall decision making process steps remain the same in risk-based decision making, namely to define the issues, examine the options and implement the resulting decision. What is different, however, is that the decision is arrived at by a structured understanding of the risk-reward balance and uncertainties. The options available will be based on one or more of the '4Ts' risk response actions: Terminate, Treat, Tolerate, Transfer as shown in Figure 1.7.

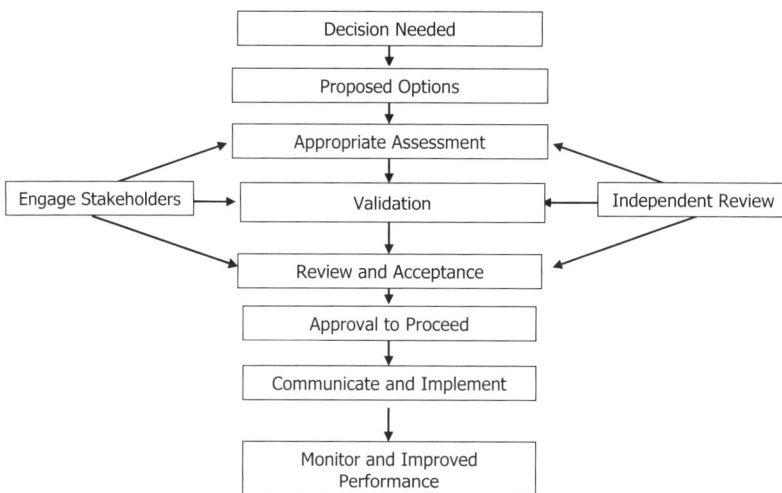

Figure 1.6   Standard process for decision making

Treat
This strategy seeks to reduce the risk probability or its impact by taking early action to reduce the occurrence of the risk to an acceptable limit. Risk mitigation may take the form of implementing new processes, undertaking more preliminary work or selecting more stable suppliers. Risk mitigation can also include changing conditions so that the probability of the risk is reduced, by adding resources or time to the programme.

Terminate
Risk termination or avoidance is changing the project plan to eliminate the risk or to protect the project objectives from its impact. Although not all risks can be eliminated, some may be avoided by taking this pre-emptive action.

4Ts

Transfer
Risk transfer is seeking to move the consequence of a risk to a third party together with ownership of the response. Transferring the risk does not eliminate it; it simple gives another party responsibility for its management. This is the most effective way of dealing with financial risk exposure and can be by a contract to another party or by payment of a premium in the case of insurance.

Tolerate
This strategy indicates that the project has decided not to change the project plan and to deal with a risk, or is unable to identify any other suitable strategy to adopt. Risk acceptance may also occur when the cost of dealing with it would not be cost effective. In this event the development of a contingency plan to execute should the identified risk occur is a natural step. Active risk tolerance may include developing a contingency plan to execute should a risk occur. Passive tolerance requires no action leaving the project team to deal with the risks as they occur.

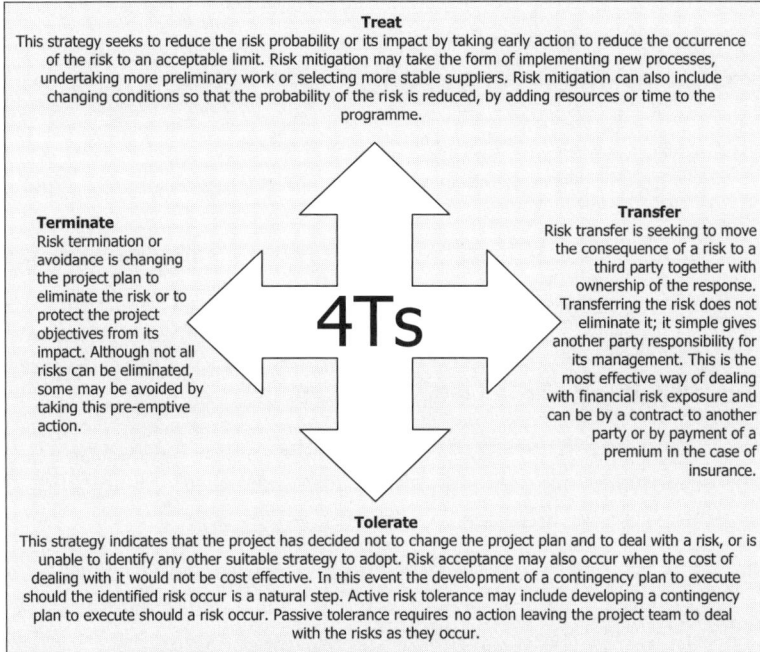

Figure 1.7   Risk response actions

Source: Figure developed from '*The Essential Management Toolbox*', S.A. Burtonshaw-Gunn (2008). © John Wiley and Sons. Reproduced with permission.

In addition to the above 4T risk response actions is the additional option of rejecting the risk if it is considered that the occurrence of the risk is so improbable that it will not be a threat to the project.

A well-designed risk response portfolio will focus not only on reducing the likelihood of a risk occurring, but should also include plans for stabilization and recovery to ensure business continuity and effective reputation management. It may also be possible to reduce the potential for financial loss by long-term fund-hedging techniques or from the use of insurance policies. Following this an evaluation of the risk response options will be required, taking into account their cost, benefits and the views of all relevant stakeholders. Whilst risk responses which are not cost-effective (that is, the value of any reduction in risk is outweighed by the cost of the control) would normally be discarded, there may be mandatory requirements imposed by internal standards or external regulatory authorities. Ultimately, a decision is made which often is clear-cut inasmuch as the proposal is clearly worthwhile or not. At other times, however, there is no clear answer, requiring further investigation of the underlying issues or a simple consensual decision. Any decision requires an assessment of whether the 'residual' risk is acceptable, given the risk appetite of the organization which, while difficult to quantify, is surprisingly well understood, if subconsciously, within most organizations.

Whilst this process is reasonably straightforward in principle, in practice there can be demanding issues to overcome, for example:

- Ensuring the options have been properly selected and defined.

- Setting assessment criteria, and objectives and their relative importance.

- Identifying risk issues and perceptions.

- Assessing the performance of options against aspects which may not be quantifiable, or which may involve judgments and perceptions that vary or are open to interpretation.

- Dealing with differences in the uncertainties of estimates, data and analyses – it may not be able to provide a fair reflection of the actual differences between the options being considered.

- Managing or avoiding hidden assumptions or biases.

In looking at risk management from a wider stance than that of the construction industry, effective risk-based decision making processes have common features regardless of the business application, and good examples can be found in the United Kingdom Offshore Operators Association (UKOOA) decision making framework, arguably the best known within the high hazard industries, and also in the Rail Safety and Standards Board guidelines on risk-based decision making, to name but two from other industries.

Risk Management and Culture

The final topic of this opening chapter is that of the topic of the 'Risk and Management Culture'. As covered earlier there are many definitions of risk management and even the ones most commonly used are often interpreted by construction companies to align to their own culture, client base and strategic aspirations as shown in Figure 1.8. However, successful project risk management requires three key attributes to be not just in place but more importantly embedded within the companies. These are, firstly that a systemic risk process is in use; secondly, that staff involved in construction projects have an appropriate knowledge and experience of risk management; and finally that there is a supporting culture to risk management from the senior management team.

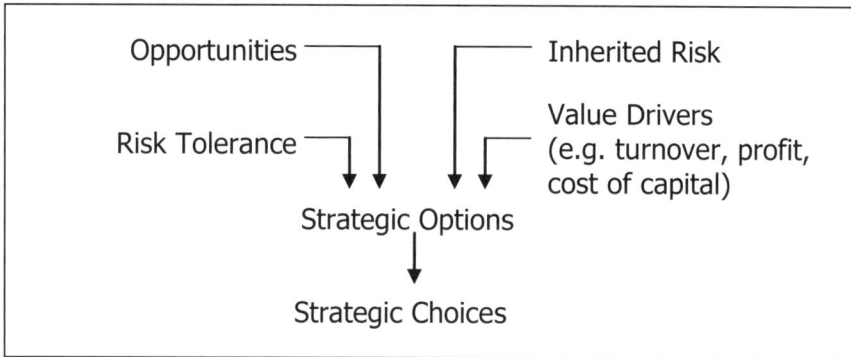

Figure 1.8   Relationship of risk on strategic management

On this final point it is stressed that building the necessary culture is rarely a simple or straightforward process as it often takes significant time, resources and commitment. Whilst the framework for an organizational culture is established and reinforced by the senior management team in their leadership style and commitment which drives process, performance and policy through measurement and audit, the key features are also derived from the staff in sharing a common set of behavioural attitudes, goals and personal attributes. Clearly these two attributes of personal behaviour and corporate culture are interlinked with each influencing the other such that an organization's corporate attitude to risk management is for many companies directly related in the staff it attracts and retains.

The final model (Figure 1.9) of this chapter illustrates how both employee and employer behaviour, values and culture play a joint role in converting the organization's desired performance targets into meeting and supporting longer-term strategic objectives. It can be used to gain an understanding and match of both sides of the equation: on one hand, from the employee viewpoint, is understanding their capabilities, level of involvement and individual values compared and linked to the employer's position. On the other hand, are organizational design, reward and corporate values, including the employers approach to managing the 'soft' issues.

In considering risk management in construction projects, such cultural knowledge will provide a better match between the individual capability and the organization's set performance targets.

Rewards
• Recognition and how people are rewarded
• Appraisal and performance reviews and
how performance is managed
• Remuneration and incentive bonus

Organization's Values
• Present and historical business
performance
• Customer focus
• Trust and respect
• Employee welfare

Organization Design
• The structure and shape of the
organization
• How work is valued
• The employer – employee relationship
and dynamic

Organizational performance measures → **Desired Behaviours** → **Supports strategic objectives**

Employee Involvement
• Communications
• Expectations
• Personal development opportunities
• Teamwork/friendships

Workforce Commitment
• Commitment to task –
accountability and trust
• Confidence in leadership skills,
communications, respect
• Sense of pride in being part of
the organization

Individual's Values
• Ethics and morals
• Belief and background
• Openness
• Attitude

Capabilities
• Skills and training
• Knowledge, experience, qualifications
• Professional recognition
• Information and data

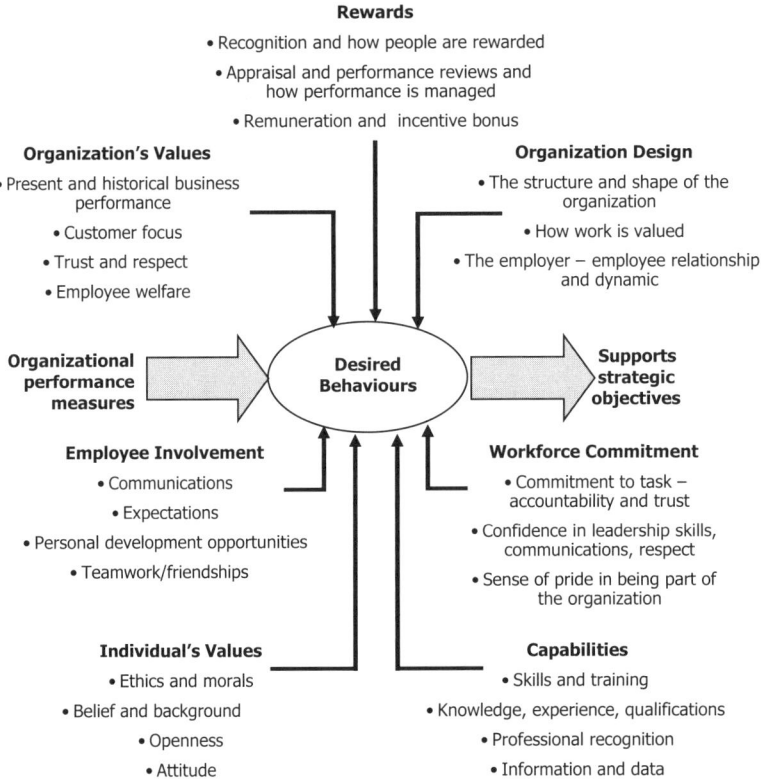

Figure 1.9   Culture and performance

Source: Figure developed from '*The Essential Management Toolbox*', S.A. Burtonshaw-Gunn (2008). © John Wiley and Sons. Reproduced with permission.

In looking at the project lifecycle, the main aims of risk management are to:

• Identify potential risks.

• Assess the probability and impact of each risk.

• Identify alternative actions that prevent the risk from happening (avoidance), or if it does happen ameliorate the impact (reduction), or provide a strategy for dealing with the accepted consequences of such a risk occurring (acceptance).

• Implement and monitor those actions that are cost effective and necessary to the successful delivery of the project objectives (NB: project objectives not project).

- Provide feedback from experiential learning to improve the risk management of future projects and to inform the training and development of project managers.

Risk management should therefore be regarded as an integral part of project management and not as an additional extra. It should be used to drive, inform and support planning and from this overview of risk management in construction it should be seen that effective risk management can:

- Anticipate and influence events before they happen by taking a pro-active approach.

- Provide knowledge and information about predicted events.

- Inform and, where possible, improve the quality of decision making, recognizing the preferred hierarchy of risk avoidance, risk reduction, risk control and risk acceptance.

- Avoid covert assumptions and false definition of risks.

- Make the project management process overt and transparent.

- Assist in the delivery of project objectives in terms of benchmarked quality, time and cost thresholds.

- Allow the development of scenario planning in the event of the identification of a high-impact risk.

- Provide improved contingency planning.

- Provide verifiable records of risk planning and risk control.

To achieve effective and efficient risk management, risk planning is required. The commonest form of risk planning is the Risk Management Plan which is described in the next chapter.

Risk Identification and Planning

Risk Classification

Before exploring the two main topics of the Identification and Planning activities detailed in this chapter it should be noted that risks can have a classification system. This system simply classifies risks in relation to their focus of action. That is the organizational level at which the risk will have the most impact. These could cover four areas and is a separate concept to, but not disconnected from, the risks within the project lifecycle. These risks areas are:

- **Project Risks** – risks within the project scope of work that could affect the delivery of the business outcome that the project is set up to deliver. In other words those risks which could affect the delivery of the project's objectives.

- **Business Risks** – risks, on the other hand, that affect the operation of the business outcome once it has been delivered by the project.

- **Environmental Risks** – risks that are external to the project environment but which nevertheless can affect the project objectives. For example, the Gulf War had a devastating effect upon gas field projects in Kuwait back in 1990.

- **External Change Risks** – risks that are beyond the immediate project environment but which could have a major impact. Frequently in contractual terms these may include *force majeure* events. However, external change risks go beyond *force majeure*, for example, because of a change in government policy or in its interpretation of a law.

Figure 2.1 illustrates how different risk management tools may be appropriate when attempting to make a business decision, depending on the level of risk and operational complexity.

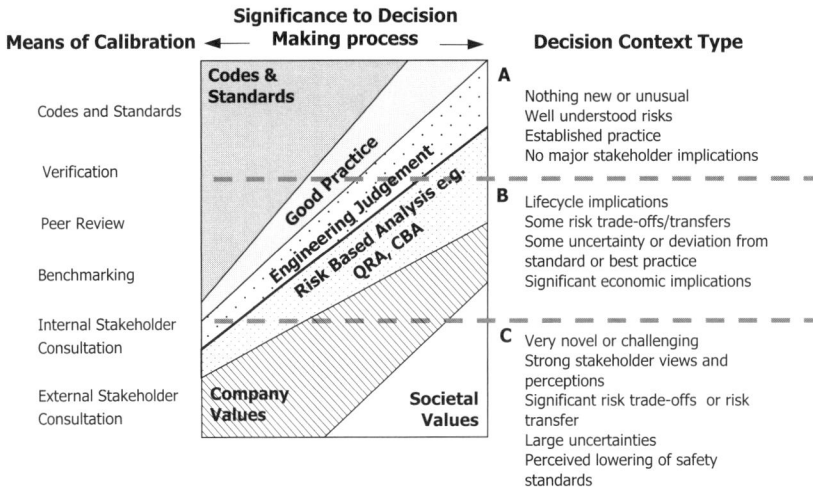

Figure 2.1   Risk management tools

Source: UKOOA – A Framework for Risk Related Decision Support, reproduced with kind permission.

Risk Identification

Risk Identification involves determining which risks might affect the project and documenting their characteristics. Participants in risk identification need to be selected on their ability to identify risks in a given technical or management area and this generally includes a number of the following:

- the project team;

- the risk management team;

- subject matter experts – for example ground conditions engineers;

- customer(s);

- end users if they are different from customer;

- other project managers with experience of similar project undertakings;

- stakeholders;

- outside experts such as public authority representatives.

Risk identification is an iterative process and often the first pass is performed by a part of the project team or by the risk management team followed by a second iteration by the entire project team with the primary stakeholders. To achieve an unbiased analysis, persons who are not involved in the project can be used to perform a final iteration. Often, simple and effective risk responses can be developed and even implemented as soon as the risks are identified. Risk identification can be undertaken using either top-down and bottom-up approaches or both. Top-down risk identification provides a rapid start to the assessment and commences with an overall view of the programme but requires an understanding of the project's mission, scope and objectives of the owner, sponsor and/or stakeholders. Output of other processes should be reviewed to identify possible risks across the project; these may include:

- product descriptions;

- schedule and cost estimates;

- resources plan;

- procurement plan;

- assumptions;

- constraints.

Alternatively bottom-up identification involves a systematic and comprehensive consideration of the project management and technical deliverables; this approach should examine the project requirements, plans, specifications, resourcing, contracts, subcontractor, subcontract characteristics and project interfaces and interactions to identify risk areas.

Typical project risks that could be encountered on a construction project are shown in Figure 2.2.

Project Management Risk	Examples
Customer	• Customer focus • Specification quality • Changing requirements
Project Management	• Planning • Resourcing • Resource capabilities • Dependencies • Stakeholders • Organization/interfaces • Communication • Constraints • Process • Transition and services

Figure 2.2   Typical project management risks

Procurement	• Planning • Vendor appraisals • Critical lead-times • Reliance on single source • Component obsolescence • Market volatility
Commercial	• Subcontractor agreement • Interpretation of Terms and Conditions
Financial	• Profit margin • Accurate cost forecasts • Payment plan • Penalty charges
Construction	• Feasibility • Technology transfer • Complexity • Dependencies • Resourcing • Special standards • Documentation requirements • Prototypes • Maturity • Manufacture • Process
Manufacturing	• Make/buy planning • Design • Production capacity • New tools/equipment requirements • Test requirements • New manufacturing or test processes, incorporating change during manufacture
System design and integration	• Systems complexity • Interfaces • Human factors • Software • Hardware
Technology	• Technology or technical approach chosen to achieve the project objective
Subcontractor capabilities	• Ability of contractors or other vendors to perform project objectives, including project management strategy and ability
Interfaces	• Working in a multi-project environment, interfacing with existing operational activities and other stakeholders
Environmental	• Environmental laws and compliances • Licences and permits
Regulatory involvement	• Involvement by any regulatory agency such as Environmental Agency, Health and Safety Executive • National, state and local governments

Figure 2.2 *Continued*

Political visibility	• Political significance or visibility to national • State or local governments • Special interest groups • The public
Number of key project participants	• Involvement by other than a primary owner for the decision making and management
Complexity	• Issues with design criteria, functional requirements • Complex design features • Breakthrough technology or existing as-built condition documents
Labour skills availability and productivity	• Adequate resources, speciality resources • Rapid labour force build-up experience and commitment • Exposure to environmental extremes
Number of locations/site access/site ownership	• Geographic dispersion • Time zone differences • Site ownership • Access issues
Funding/cost sharing	• Project duration • Involvement/funding by other parties, and stability of monetary inputs
Magnitude/type of contamination	• Presence of hazardous or mixed waste
Quality requirements	• Requirements for precision work or other QA requirements; types of QA methods
Site	• Ground conditions • Flood plain • Contaminated ground • Archaeological finds environmental protection • Protection of animals, birds, flora and fauna
Public involvement	• Citizen interest or involvement, rights of way

Figure 2.2   *Concluded*

In looking at the list in Figure 2.2 it would be fair to assume that projects are increasing in technical complexity and the number and range of risks that need to be addressed. As such, due to this level of complexity, there is a corresponding increase in the risk of not meeting the success criteria, as established in the concept and planning phase with the client of the project and also related to the three key project management variables. Historically, project decision making has been heavily biased toward meeting the cost and schedule goals without the same level of thought applied to the consequences of the project's technical objective(s). This has been the legacy of the Earned Value performance measurement approach to project management, which measures success primarily by concentrating on the two elements where a preponderance of the known data can be measured.

Tools and Techniques for Risk Identification

There are a number of tools and techniques available for use in risk identification, these are described in the Figure 2.3 below:

Documentation review	Performing a structured review of high level and detailed project plans and the prior project assumptions and any other information is generally the initial step taken by project teams.
Assumptions analysis	Every project is conceived and developed based on a set of hypotheses, scenarios or assumptions validity. Assumptions analysis is a technique that explores that assumption's validity. It identifies risks to the project from inaccuracy, inconsistency or incompleteness of assumptions.
Diagramming techniques	Diagramming techniques may include cause and effect diagram (also known as Ishikawa or fishbone diagrams). These are useful for identifying causes of risks. (See also Figure 3.5 in Chapter 3). Systems or process flow charts – these show how various elements of a system interrelate and the mechanism of causation.
Checklists	Checklists for risk identification can be developed based on historical information and knowledge that has been accumulated from previous similar projects and from other sources of information. One advantage is that risk identification for this method is quick and simple. However, a disadvantage is that it is impossible to build an exhaustive checklist of risks, and the user may be effectively limited to the categories in the list. Care should be taken to explore items that do not appear on a standard checklist to determine if they seem relevant to the specific project. The checklist should itemise all types of possible risks to the project. It is important to review the checklist as a formal step of every project-closing procedure to improve the list of potential risks and the description of risks for subsequent projects, (See also Chapter 4). It should be noted that checklists are seldom exhaustive but can help to ensure that the most common key areas of project risk are considered. They are particularly useful as 'prompts' to facilitate brainstorming. A sample checklist is provided as Appendix 1 at the end of this chapter.

Figure 2.3   Tools and techniques for risk identification

Information-gathering techniques	Examples of information-gathering techniques used in risk identification can include:
	• Brainstorming sessions – this is probably the most frequently used risk identification technique. The goal is to obtain a comprehensive list of risks that can be addressed later in the qualitative and quantitative risk analysis processes. The project team usually 'brainstorms', although a multidisciplinary set of experts can also use this technique. Under the leadership of a facilitator, ideas are generated about project risks which are discussed and agreed as a set of risks. Sources of risk are identified in broad terms and displayed for all to examine during the meeting. This method is popular, can cover a lot quickly and encourages the participants to think wider than the project boundaries. The key point is not to discount any suggestion but to initially capture everything. Consideration of which risks to accept or to manage, comes later by debate and agreement of which to address or which to discount. The disadvantage of this technique is that it can be dominated by strong characters that may, whether intentionally or not, prevent others from contributing. • The Delphi technique is a variation of brainstorming where contributions are made to a coordinator. It is a good method to reach a consensus of experts on a subject and is well suited to the identification of project risk. In practice project risk experts are identified but participate anonymously. A facilitator uses a questionnaire to solicit ideas about the important project risks and after the responses are returned they are then circulated to the experts for further comment. Consensus on the main project risks may be reached by a few rounds of this process. Whilst this technique used to be a slow process, nowadays web-based tools allow this to be undertaken much faster. An advantage of the Delphi technique is that it helps to reduce bias in the data and keeps any single expert participant from having undue influence on the outcome. • Interviewing can allow risks to be identified by interviews of experienced project managers or subject-matter experts. The person responsible for risk identification identifies the appropriate individuals, briefs them on the project and provides information such as the WBS and a list of any assumptions made. The interviewees identify risks to the project based on their experience, the project information and any other sources that may be thought to be useful. This method has two disadvantages; firstly it takes longer than brainstorming and secondly experts may underestimate project risks in their own areas of expertise. • SWOT analysis covering Strengths, Weaknesses, Opportunities and Threats ensures examination of the project from each of the SWOT perspectives to increase the breadth of the risks considered.

Figure 2.3   *Continued*

	• The use of a detailed checklist of risks can be considered and is regarded as a fast method of risk identification although may prevent specific project risk consideration by assuming that everything is covered on this checklist. It can also make it difficult to think of risks outside the list.
Observation or Learning from experience	Make use of near neighbour comparisons of similar projects, locations, suppliers, customer etc. Close examination of a current systems or project may help to identify risk and may also be inherent to a new project.

Figure 2.3   *Concluded*

After identifying the risks that may affect a project these can then be organized into risk categories to reflect common sources of risk for the industry or application area. For construction projects, typical categories will include:

- Technical, quality or performance risks such as reliance on unproven or complex technology, unrealistic performance goals, changes to the technology used or the industry standards during the project.

- Project management risks such as poor allocation of time and resources, inadequate quality of the project plan, poor use of project management disciplines.

- Organization risks such as costs, time and scope objectives that are internally inconsistent, lack of prioritization of projects, inadequacy or interruption of funding and resource conflicts with other projects in the organization.

- External risks such as changing legal or regulatory environment, labour issues, changing owner priorities, country risk and weather. *Force majeure* risks such as earthquakes, floods and civil unrest for example generally require disaster recovery actions rather than those of risk management and therefore are generally excluded from the vast majority of construction project risk plans.

Having used a number of tools and techniques to identify the risks for a project it is important that the risk statement is correctly phrased to reduce any misunderstandings; as such it is recommended that this should follow the condition-cause-consequence approach, two examples of which are shown below:

'There is a risk that the customer will be unable to specify the internal fit-out requirements in a timely fashion caused by their lack of experience in procuring this type of equipment resulting in delayed payment, project overrun and delayed initiation of support contracts.'

'There is a risk that the client will wish to bring forward the completion date for the project and cause the contractor to execute the work by additional shift working or the allocation of more resources resulting in an overall cost increase from that contractually agreed.'

Risk Planning

Risk Planning is the process of deciding how to approach and plan for the risk management activities of a project. This is an important step to ensure that the level, type and visibility of risk management are commensurate with both the risk and importance of the project to the organization.

The inputs to the risk management planning activity of the project are shown in Figure 2.4 and are described below:

- Some organizations may have predefined risk management policies and approaches to risk analysis and responses that have to be tailored to a particular project.

- Predefined roles and responsibilities and authority levels for decision making will influence the risk planning activities.

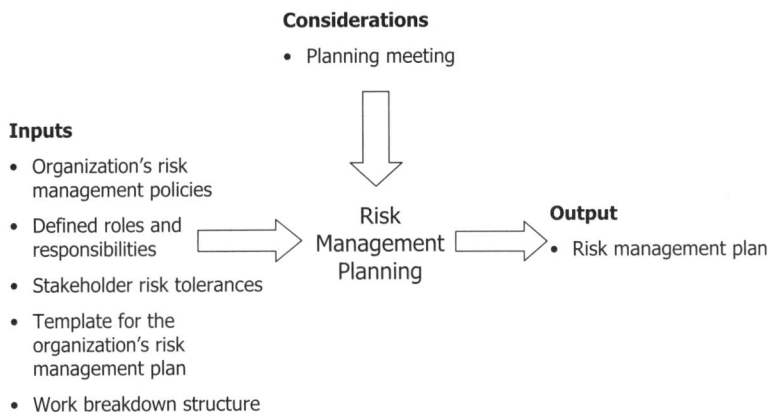

Considerations
- Planning meeting

Inputs
- Organization's risk management policies
- Defined roles and responsibilities
- Stakeholder risk tolerances
- Template for the organization's risk management plan
- Work breakdown structure

Risk Management Planning

Output
- Risk management plan

Figure 2.4   Elements of risk management planning

- In considering stakeholder risk tolerances, different organizations and different individuals will have different risk tolerance levels. These may be expressed in policy statements or revealed in personal actions and attitudes. Chapter 6 also discusses these risks with respect to financial risk groupings.

- Some organizations have developed templates (or a pro-forma standard) on which to base a risk management plan for use by the project team. In time the organization is likely to improve the template based on its application and usefulness in its projects for the benefit of future projects.

The main tool for the risk management planning is the planning meetings, at which the project team develop the risk management plan. Attendees usually include the project manager, the project team leaders, anyone in the organization with responsibility to manage the risk planning and execution activities, key stakeholders and others, as and when needed. These meetings can make use of the organization's risk management templates and other inputs as appropriate. The primary output from the meeting should be an agreed management approach to risks that the project will face and a formal risk management plan which will detail how risk identification, qualitative and quantitative analysis, response planning, monitoring and control will be undertaken during the project's lifecycle. Importantly the risk management plan does not address responses to individual risks as this is accomplished in the risk response plan. As such the risk management plan may include:

- A methodology to define the approaches, tools and data sources that may be used to perform risk management of the project. Different types of assessments may be appropriate depending upon the stage of the project, the amount of information available and the flexibility remaining in risk management.

- The detail of the roles and responsibilities and the lead, support and risk management team membership for each type of action in the risk management plan. Risk management teams drawn from outside of the project may be able to perform a more independent, unbiased risk analysis of the project than those from the direct project team.

- Establishing a budget for risk management for the project.

- Defining how often the risk management process will be performed throughout the project life-cycle. Results should be developed early enough to affect decisions, indeed these decisions should be periodically reviewed during the project execution phase. Typically a project manager will review the risks on a monthly basis; sometimes this review is more frequent where the risks have a significant financial or programme impact.

- Detailing the scoring and implementation methods appropriate for the type and timing of the qualitative and quantitative risk analysis to be performed. It is good practice for the methods and scoring to be determined in advance to ensure a consistent approach.

- A description of the threshold criteria for risks that will be acted upon, by whom, and in what manner. The project sponsor and customer may have different thresholds and the acceptable threshold criteria will form the target against which the project team can measure the effectiveness of the risk response plan execution.

- A description of the content and format of the risk response plan and how the results of the risk management processes will be documented, analyzed and communicated to the project team, internal and external stakeholders, sponsors and other interested parties as appropriate.

Many organizations employ a graded approach to risk planning which may have a number of criteria for determining the application of this approach such as financial value, complexity, visibility, strategic risk and so on. The risk planning provides a formalized and documented method or technique to determine the graded approach allowing all interested parties to have an opportunity to identify and assess the impact of a wide variety of potentially adverse risks on the project's technical objectives. Following this, cost and schedule adjustments can then be incorporated to present a more realistically achievable estimate of the resources necessary to achieve a successful project. The resultant baseline then becomes a much more useful tool in managing the project and its expectations. Indeed, without this planning, the confidence in completing the project to its described success criteria must be necessarily low.

Within the risk planning stage it is important to document the overall risk management process so that all parties are aware of the implementation and its ongoing review arrangements. The contents of a typical risk management plan are shown in Figure 2.5.

Title	Details
Introduction	Project/Product Overview: • Summary of requirement. • Critical success factors. • Project life-cycle.
Control of Plan	Overview: • Review and reissue frequency.
Scope and Objectives	Scope of work: • Scope, complexity and scale of project. • Initial assessment of difficulty, scale, precedence, impact of failure. Objectives: • Deliverables: e.g. risk register, report and mitigation plans. • Reporting requirements for project team, main contractor and subcontractor(s).
Identification Strategy	Identification: • Describe the identification process: how the risks are identified e.g. 'brainstorming', checklists etc. • The discipline of describing risks. • How ownership is established and recorded e.g. Risk Register. • How new risks are identified and mitigated risks retired. • When risks are identified and at what level. Allocation: • Describe the process to assign and apportion risk to other stakeholders e.g. subcontractors, partners, other areas of the organization.
Assessment Strategy	Analysis (qualitative): • Describe the method used to establish post mitigation criticality scores from probability and impact assessments. Evaluation (quantitative): • Describe the approach to be taken to evaluate collective cost, project and performance exposures.

Figure 2.5   A typical risk management plan

	Response strategy: • Describe how performance, cost and effectiveness benefits will be calculated to determine courses of action (option selection). • Describe how pre-emptive risk action will be fully integrated into the overall project programme. • Describe how corrective fallback plans will be incorporated into the overall project programme. • Describe how the risks are to be treated, transferred, tolerated and terminated. • Describe fallback measures to recover in the event of risk occurrence. • Identify how associated costs and effectiveness of action will be established and recorded. • Identify how contingency will be released to support effective mitigation action on the occurrence of risk.
Process Management	Risk Management process: • Process objectives. • Outline process description, clearly identifying the supporting audits and reviews including contract risk reviews and technical risk reviews. Contingency Management: • Describe the principles and methods for determining the correct overall level of project contingency. • Allocating technical and managerial contingencies and movements between the two. • Authorization of pre-emptive mitigation spend. • Authorization of the release of contingency funds to support corrective risk mitigation. • Authorization of risk retirement. Risk Management tools: • Identifying the tools and methods being used to support the risk assessment, support requirements and maintenance responsibilities. Risk Register control: • Describe the process for maintaining the register, indicate how items are to be entered, updated and deleted. • How associated mitigation/promotion actions/plans/ programmes and events will be recorded and reviewed. • Detail where the register will be kept, how it will be accessed and by whom.

Figure 2.5   *Continued*

	Risk Reporting:
	• Describe the reports to be generated from the process including details of: what reports will be generated; what the reporting cycle will be; and at what level e.g. top ten most significant risks. • Describe the process for identification of new risks, deletion and retirement of old risks.
Organization	Project responsibilities:
	• Identify clearly the role, authority and responsibility with respect to risk and mitigation actions of the project manager/director, risk manager, risk owner, risk actionee and the key project team members. • Provide a list of people who will have responsibility for regular review of the risks, detailing their roles. • Provide a summary of internal and external parties involvement with the risk management process e.g. customer, subcontractor, suppliers, user, government agencies.
	Functional responsibilities:
	• Define the main functional interfaces (customer/supplier) between the various project groups or areas. • Identify how risk and associated contingency will be allocated to the functional areas and managed across the organizational interfaces.
Programme	Mobilization programme:
	• The timing and initiation of the Project Risk Management process from initial identification and assessment programme and responsibilities for and selection of supporting systems and tools.
Metrics	What key performance indicators, if any, are to be collected. Typical key performance indicators include, for example:
	• Regularity/coverage of reviews. • Total cost exposure. • Interim cost exposure (vs. payment milestones). • Overall schedule exposure (completion). • Interim schedule exposure (vs. delivery milestones). • Performance exposure (compliance or acceptance metric). • Number of risks impacting. • Total number of risks over time. • Mitigation spend to exposure reduction ratio. • Risk trends. • Action plan progress.

Figure 2.5   Concluded

It should be noted that on small projects the risk management plan may be incorporated into an overall project plan, where this is appropriate the contents and rationale of a typical project plan is shown in Figure 2.6

The terms 'Probability' and 'Impact' have already been previously mentioned, however, on a project it is important that all parties have an agreed shared understanding of these terms and one of the early tasks in the project will be to agree the criteria of these for the specific project depending on the value, timescale pressure and overall project requirements. In all cases the definitions can be mapped on a scale ranging from 'Very Low' to 'Very High' as shown in Figure 2.7.

Why ?

- To demonstrate that the business needs are understood including those of the customer for the project, its viability and that business policies and strategies are satisfied by the project.

What ?

- The Project Plan needs to show clearly the defined project objectives: a statement of requirements, scope of work, key performance indicators, priorities and the project's time, cost and quality targets.

How and When ?

- The plan reports on the agreed strategy and approach with work breakdown structure, organizational structure and resourcing plan. It should detail the defined roles and responsibilities, the project's payment schedule, cash flow forecast, vendor strategy and plans together with any contingency plans as they are developed and the Project Plan updated.

How Well ?

- Suitable control and reporting systems for time and budget, risk management plan, quality assurance plan, vendor compliance, claims and variations, change management system need to be documented in the plan.

What if ?

- As part of the project considerations a section on risk assessment and contingency plans describing actions to avoid and mitigate, responsibility for monitoring threats and contingency planning should be included.

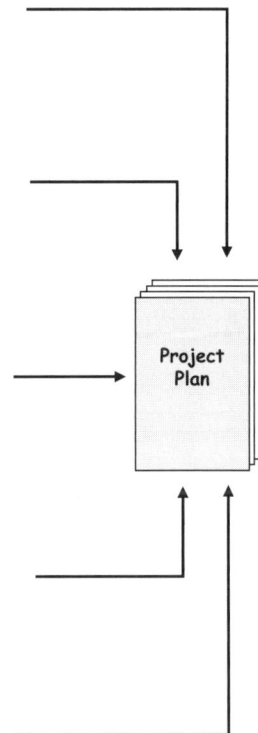

Project Plan

Figure 2.6   Project plan rationale

Source: Figure developed from '*The Essential Management Toolbox*', S.A. Burtonshaw-Gunn (2008). © John Wiley and Sons. Reproduced with permission.

Probability Definitions:	
Qualitative guidance: ... the risk occurrence	Quantitative Guidelines
Very High: is very likely	More than 80 per cent
High: is more likely than not	Between 51 to per cent
Medium: is less likely than not	Between 20 to 50 per cent
Low: is very unlikely	Between 5 to 20 per cent
Very Low: is extremely unlikely	Not as much as 5 per cent

Cost Impact	
Very High	... jeopardises the business budget due to a substantial provision for consequential damages.
High	... endangers the project financial viability i.e. pose a significant margin threat so that substantial replanning requires the reassignment of business's resources sufficient to harm other projects or cause considerable consequential delay costs which have cost implications.
Medium	... requires significant rework or the selection of alternative subcontractors or cause significant consequential delay costs and adjustment of the risk-margin balance.
Low	... requires additional or higher-rated resources or a change to alternative suppliers or cause some consequential delay costs.
Very Low	... requires adjustment of resource profiles, resulting in non-optimum efficiency.

Schedule Impact	
Very High	... causes significant programme completion non-compliance, unrecoverable by replanning.
High	... causes unavoidable programme milestone non-compliance for which replanning is required.
Medium	... causes overrun and certain work for which re-planning is required.
Low	... requires probable rework.
Very Low	... requires some rescheduling of tasks.

Performance Impact	
Very High	... results in a serious non-compliance or degradation in performance sufficient to terminate the project.
High	... causes a problem or degradation in performance so significant that a substantial non-compliance is unavoidable.
Medium	... causes partial compliance that requires concessions or difficult negotiations.
Low	... causes a problem that requires trade-off studies or negotiations to resolve.
Very Low	... requires some adjustment in the solution.

Figure 2.7   Probability and impact definitions

On the basis of the agreed definitions in the above tables, the project can then use these to construct a probability versus impact matrix as shown in Figure 2.8 where the critically scores determines the risk significance and allows prioritization of the project tasks in reduction to one relation to one another, the risk reporting thresholds and establishing a tolerance range. The table of impact definitions (Figure 2.9) also reflects the severity of the risk on the project. Impact can be ordinal or cardinal, depending upon the culture of the organization conducting the analysis. Ordinal scales are simply rank-order values such as Low, Medium and High. Cardinal scales assign values to these impacts and there scales are usually linear, such a 1, 3, 5, 7, and so on but may be non-linear 0.5, 1, 2, 4, 8, reflecting the organization's desire to avoid high-impact risks. The intent of both approaches is to assign a relative value to the impact on project objectives should the risk in question occur.

		Probability				
		Very Low	Low	Medium	High	Very High
Impact	Very High	4	6	8	10	12
	High	3	5	7	9	11
	Medium	2	4	6	8	10
	Low	1	3	5	7	9

Figure 2.8   Probability/impact matrix

	Impact Definitions				
Project	Very Low	Low	Medium	High	Very High
Objective	0.05	0.1	0.2	0.4	0.8
Cost	Insignificant cost increase.	<5% cost increase.	5–10% cost increase.	10–20% cost increase.	> 20% cost increase.
Programme	Insignificant Schedule slippage.	Schedule slippage <5%.	Overall project slippage 5–10%.	Overall project slippage 10–20%.	Overall project slippage >20%.
Scope	Scope decreases, barely noticeable.	Minor areas of scope are affected.	Major areas of scope are affected.	Scope reduction unacceptable to client.	Project end. Item is effectively useless.
Quality	Quality degradation barely noticeable.	Only very demanding applications are affected.	Quality reduction requires client approval.	Quality reduction is unacceptable to client.	Project end. Project is effectively unusable.

Figure 2.9   Table of impact definitions

Well-defined scales, whether ordinal or cardinal, can be developed using the definitions agreed by the organization and the appropriate project teams. These definitions improve the quality of the data and allow the process to be more repeatable.

The final table of this chapter (Figure 2.10) is another probability impact matrix which may be used to assign risk ratings (low, medium, high) to risks or conditions based on combining probability and impact scales. Risks with high probability and impact are likely to require further analysis, including quantification and aggressive risk management. The risk rating is accomplished using a matrix and risk scales for each risk.

A risk's probability scale naturally falls between 0.0 (no probability) to 1.0 (absolute certainty). Assessing risk probability may be difficult because expert judgement is used often on unique construction projects but without the benefit of historical data. An ordinal scale representing relative probability values from very unlikely to almost certain could be used in these circumstances. Alternatively specific probabilities could be assigned by using a general scale, (for example 1, 3, 5, 7, 9).

The above approach to assessing risks can be used to define risk thresholds for different risk categories. Effective setting of a level of tolerance supports effective and efficient project management and risk tolerance levels can also drive the reporting of risks by exception and escalation. In considering risk tolerance it is natural that there will always be some risks that the organization is not prepared to take and on every project the organization's stakeholders will always influence of the limits of risk acceptability.

		Impact				
		0.05	0.10	0.20	0.40	0.80
Probability	0.9	0.05	0.09	0.18	0.36	0.72
	0.7	0.04	0.07	0.14	0.28	0.56
	0.5	0.04	0.05	0.10	0.20	0.40
	0.3	0.02	0.03	0.06	0.12	0.24
	0.1	0.01	0.01	0.02	0.04	0.08

Figure 2.10   Probability impact matrix (Risk score = P × I)

Appendix 1
Example Checklist for Risk Identification

Category	Yes	No
Technology		
Does this project involve new technology?		
Does the project involve any unknown or unclear technology?		
Does the project offer new applications for existing technology?		
Does the project include for modernizing advanced technology in its existing application?		
Time		
Are there any project schedule uncertainties or restraints that may impact project completion or milestone dates?		
Are there any long lead times (LLT) for procurements that may affect critical path or milestone completion?		
Subcontract Capabilities		
Are there potential transportation or infrastructure impacts?		
Interfaces		
Does the project significantly interface with the current operational facility?		
Safety		
Is there significant contamination potential?		
Does the new design basis give us any concern regarding potential for accidents or prompt any other safety questions?		
Does any hazardous material have to be removed?		
Environmental		
Is an environmental assessment or environmental impact statement required?		
Is there any potential for releases or additional releases?		
Any environmental permits or licences required?		
Regulatory Involvement		
Do any state regulators need to be involved in any project decision?		
Is the Environmental Protection Agency involved in any project decision?		
Are the Health and Safety Executive involvement in the project?		
Political Visibility		
Does the profile of project attract political interest?		
Does the project contribute to trading/diplomatic position between counties?		
Number of Key Participants		
Will there be more than one prime contractor performing the work?		
Can we use our preferred suppliers?		
Do we have to use the client's recommended suppliers?		

Complexity		
Are the functional requirements undefined or unclear?		
Is the design criteria undefined or unclear?		
Does the project have any complex or novel design feature(s)?		
Will it be difficult to functionally test?		
Are the existing or as-built conditions to be documented?		
Labour Skills Availability/Productivity		
Are adequate and timely resources available?		
Are speciality resources required?		
Is a rapid labour build-up required?		
Will labour be exposed to environmental extremes (heat, cold, etc.)?		
Will any project work be performed in a radiologically controlled zone, confined spaces and under a Permit to Work (PTW) system etc?		
Number of Locations/Site Access/Site Ownership		
Will the project work be performed in more than one physical location (area, sites, buildings, etc.)?		
Who owns the site?		
Are infrastructure improvements required?		
Is there sufficient services available – electricity, water, etc. for construction and occupation phases?		
Are these services local to the site?		
Funding/ Cost Sharing		
Is project duration greater than 2 years and how will this impact of funding?		
Are other government/corporate/foreign entities providing funding?		
Magnitude/Type of Construction		
Has waste present been characterised?		
Is hazardous or low-level waste present?		
Is high-level or mixed waste present?		
Quality Requirement		
Is precision work required?		
Does work need to be undertaken in special areas e.g. clean conditions, food preparation areas?		
Is rework expected to the nature of project tolerances?		
Public Involvement		
Is there any need for public consultation?		
Is this reflected in programme?		
Programme		
Will the execution require day/night work or shift working?		
How do national or local holiday arrangements interface with the work?		
Other		

Qualitative Risk Analysis and Quantitative Risk Evaluation

Qualitative Risk Analysis

Qualitative Risk Analysis at its simplest involves only a description of the obvious project risks; in some circumstances risk identification may be all the risk analysis that is required, in other cases more in-depth analysis will be warranted. Risk analysis is qualitative and comprises qualifying and prioritizing the risks that have been identified in terms of likelihood and impact on the construction project undertaking. Each risk must be allocated to an owner who has an understanding of the risk, an interest in its resolution and the ability to lead the analysis and all further activities on managing it. For very large projects this may require the appointment of a full-time risk manager who is able to devote all or most of their time to ensuring that a strategy exists for all project risks and that it is regularly reviewed. Qualitative risk analysis is one process of assessing the impact and likelihood on the identified risks. This process prioritizes the risks according to their potential effect on the project objectives and is one way to determine the importance of addressing specific risks and guiding appropriate risk responses. The time criticality of risk-related actions may also magnify the importance of a risk and how this is addressed. An evaluation of the available information on a regular basis can also help to modify the assessment of the risk.

Qualitative risk analysis requires that the probability and consequences of the risks be evaluated using established qualitative analysis methods and tools. When qualitative analysis is repeated any trends in the results can indicate the need for more or less risk management action. Use of these tools helps to correct biases that are often present in a project plan as described in the previous chapter. Qualitative risk analysis should be reviewed during the project's lifecycle to remain current with any change in the project risks. This process can itself lead to quantitative risk evaluation. A qualitative risk analysis process is shown in Figure 3.1.

Project Considerations

- Risk probability and impact
- Probability/impact risk rating matrix
- Project assumptions testing
- Data precision rating

Inputs

- Risk management plan
- Identified risks
- Project status
- Project type
- Data precision
- Scales of probability and impact
- Assumptions

Risk Analysis Process

Outputs

- Overall risk ranking for the project
- List of prioritized risks
- List of risks for additional analysis and management
- Trends in qualitative risk analysis results

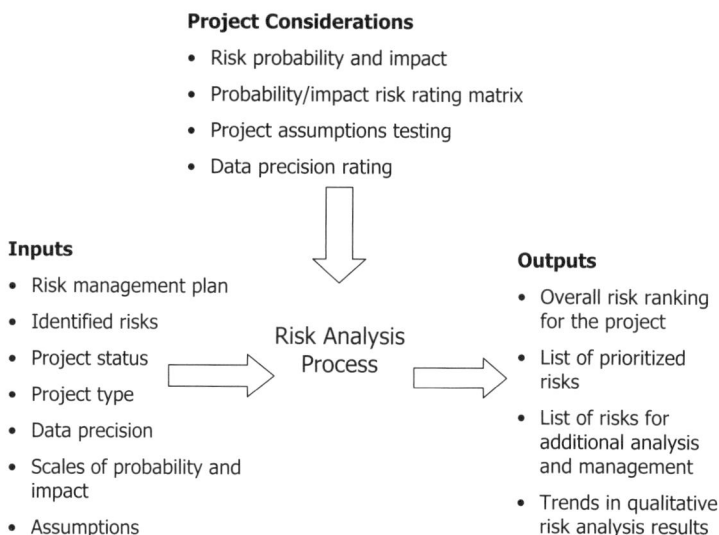

Figure 3.1   Qualitative risk analysis process

In Figure 3.1 the main inputs include:

- **Identified risks:**   Those discovered during the risk identification process are evaluated along with their potential impacts on the project.

- **Project status:**   The uncertainty of a risk often depends on the project's progress through its lifecycle. Early in the project many risks have not immerged, the design for the project is immature, and changes often occur as part of the development and feasibility stages making it likely that more risks will be discovered over time.

- **Project type:**   Projects of a common or recurrent type tend to have a better understood probability of occurrence of risk events and their consequences. Projects using state-of-the-art, innovative, leading edge construction processes or technology, or highly-complex projects typically have more uncertainty.

- **Data precision:**   The term 'precision' describes the extent to which a risk is known and understood. It measures the extent of data available as well as the reliability and validity of the data. As part of this process the source of the data that was used to identify the risk must also be evaluated.

- **Scales of probability and impact:** These scales are to be used in assessing the two key dimensions of risk and are described in more detail later in this chapter.

- **Assumptions:** Any assumptions identified during the risk identification process are to be evaluated as potential project risks.

Tools and techniques for qualitative risk analysis include:

- risk probability and impact;

- probability impact matrix;

- Ishikawa (Fishbone diagrams);

- fault trees; and

- Failure Mode and Effect Analysis (FMEA).

The two elements of probability and impact are applied only to specific risks events – not the overall project. Analysis of risks using probability and consequences helps the project team or risk manager to identify those risks that should be managed more closely. In looking at a probability/impact matrix it should be noted that depending on the uniqueness of the actual project it may be necessary to produce a series of different severity matrices against the project's criteria for time, cost and quality (or performance) in addition to an understanding of the relative priority of these factors for the specific project. Figure 3.2 below provides an example of a project time/risk matrix.

Figure 3.2 shows the combined assessment of probability and impact to produce severity rating; this can be combined with a 'traffic light' coding to easily identify safe, problem and danger zones of risk.

		Probability		
		Low 0–20%	Medium 21–50%	High 51–100%
Impact (Time)	High: 1–3 months	3	3	4
	Medium: 1–4 weeks	2	3	3
	Low: 1–5 days	1	2	3

Figure 3.2   Risk severity analysis

Ranking the risks may indicate the overall risk position of a project relative to other projects by comparing the risk scores. On this basis, the data from Figure 3.3 can be used to assign personnel or other resources to projects with different risk ranking and may lead to a cost benefit analysis decision about the project, or indeed to provide a view on more strategic decisions covering project initiation, continuation or cancellation for example.

Drawing on the Figures 3.2 and 3.3 it may be appropriate for a project or organization to combine the matrices as shown in Figure 3.4. This 6×5 severity/probability matrix is one of the most commonly used, though matrices may be encountered ranging up to 10×10 where projects require a finer level of risk sensitivity.

Risk Identification	Cost Impact	Schedule Impact	Performance Impact	Criticality Score
Risk 1	L	L	L	L
Risk 2	H	H	H	H
Risk 3	H	L	L	M
Risk 4	VL	VL	H	H

Figure 3.3   Aggregating criticality scores

The matrix shown in Figure 3.4 is a 'reactive' matrix which uses historical evidence to provide guidance on the frequency term.

The basis of the risk matrix is always the same – contrasting the frequency and consequence variables. As such it is important that the scoring mechanism used is decided in advance and applied consistently across the whole project. Naturally controls should be in place to prevent the event from happening or to minimize its probability, and also to limit the consequences should the event occur or to recover from the situation and return, as soon as possible, to normal operations. As part of the risk evaluation and management process, controls are identified for each risk scenario and their effectiveness evaluated. Some types of control are more effective than others in reducing risk; these are listed below in order of increasing control effectiveness:

- Personal protective equipment – is it available and used? Is it fit for purpose?

- Administration – do the organization's procedures need to change? Are the project team able to change them? Can the organization provide more training?

- Separate – can time, distance, shields or guards be used to give the project protection from the hazard?

- Engineer – can a less hazardous design be developed? Can the equipment be modified to be made safer?

- Substitute – can the hazard be replaced with a less harmful one?

- Eliminate – can the hazard be removed all together? Is there a better way?

It should be noted that the cost of risk management is what is required to absorb the consequences or manage this from occurring. The actions on mitigation – that is to reduce the probability and/or impact before the event – can be divided into two areas: pre- and post-mitigation. Pre-mitigation involves establishing a fallback plan which can be actioned to resolve an issue; this is likely to be an alternative strategy for the achievement of an objective that requires funding to be released from contingency allowance. Post-mitigation, commonly referred to within the construction industry as Action Planning, is a strategy for the mitigation of a risk.

		Consequence			Increasing Probability				
Severity	People	Assets	Environment	Reputation	A: Never heard of in industry	B: Has occurred in industry	C: Has occurred in industry	D: Occurs several times per year in company	E: Occurs several times per year at location
0	No injury	No damage	No effect	No impact					
1	Slight injury	Slight damage	Slight effect	Slight impact	Manage for continuous improvement				
2	Minor injury	Minor damage	Limited effect	Limited impact					
3	Major injury	Localised damage	Localised effect	Considerable impact			Incorporate risk-		
4	1–3 fatalities	Major damage	Major effect	National impact		reduction measures	Intolerable – immediate corrective action		
5	Multiple fatalities	Extensive damage	Massive effect	International impact					

Figure 3.4   **Example combined risk matrix**

One final characteristic of qualitative analysis is to evaluate risk manageability and to understand the extent to which a risk can be affected. This can offer opportunities to address the easy to deal with risks – *'the low hanging fruit'* – by dealing with the causes of controllable known risks.

In looking at qualitative analysis a number of tools that may be used have been mentioned; these are now shown below:

- The 'Fishbone' cause and effect diagram was devised by Kaoru Ishikawa, who pioneered quality management processes in the Kawasaki shipyards. The cause and effect diagram explores all the possible or actual causes (or inputs) resulting in a single effect (or output) and can be used for problem solving and to examine causes of risk. An example Ishikawa diagram is shown as Figure 3.5

- Fault Tree Analysis (FTA) is a method commonly used by reliability and safety engineers to analyze fault scenarios in design and construction. Whilst fault tree and Ishakawa diagrams start by looking at the effect and then working backwards to identify the risks; in project risk management however it is more usual to conduct this from the opposite viewpoint by first listing all the possible risks and then assessing the probable effects.

- FMEA has also been imported into project risk management from reliability and quality engineering. This method starts by considering the risk events and then proceeds to predict all their possible effects in a chart form shown in Figure 3.6.

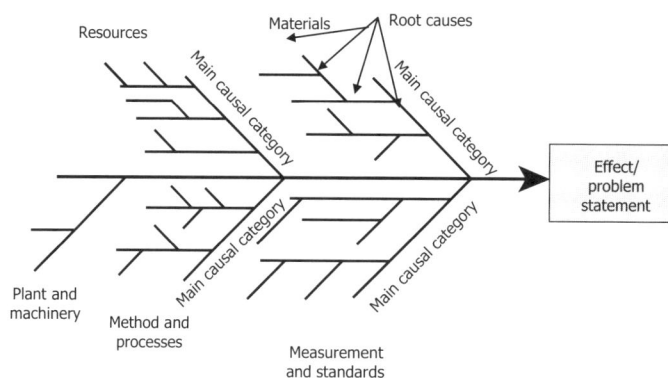

Figure 3.5   Ishikawa diagram

Source: From '*The Essential Management Toolbox*', S.A. Burtonshaw-Gunn (2008). © John Wiley and Sons. Reproduced with permission.

Item	Failure mode	Cause of failure	Effect	Remedy: recommend action
1				
2				

Figure 3.6   A typical FMEA table

Quantitative Risk Evaluation

Quantitative Risk Analysis generally follows on from the qualitative risk analysis. The quantitative risk analysis process aims to numerically analyze the probability of each risk and its consequences on the project objectives as well as the extent of overall project risk. This process uses such techniques as 'Monte Carlo' simulation and decision theory to:

- determine the probability of achieving a specific project objective;

- quantify the risk exposure for the project and determine the size of cost and schedule contingency reserves that may be needed;

- identify the risks which require the most attention by quantifying their relative contributions to project risk;

- identify realistic and achievable costs, schedule, or scope of work targets.

Quantitative risk analysis requires risk identification after which both qualitative and quantitative risk analysis processes can be used separately or together. Considerations of time and budget availability and the need for both types of analysis statements about risk and impacts will determine which method(s) to use. Trends on the results when quantitative analysis is repeated can indicate the need for more or less management action.

The inputs to the quantitative risk analysis are shown in Figure 3.7.

There are many tools available for the identification and evaluation of risks and risk controls, ranging from experience-based judgement, checklists and risk matrices, to specialist review and analysis techniques as discussed in the previous chapter. The most appropriate tool depends on the operation complexity and level of risk, and the ease of use and form of output. These are used as inputs to the quantitative risk analysis process.

Considerations

- Interviewing
- Sensitivity analysis
- Decision tree analysis
- Simulation

Inputs

- Risk management plan
- Identified risks
- List of prioritized risks
- List of risk for additional analysis and management
- Historical information
- Expert judgement
- Other planning inputs

Quantitative
Risk Analysis
Process

Outputs

- Prioritized list of quantified risks
- Probabilistic analysis of the project
- Probability of achieving the cost and time objectives
- Trends in quantitative risk analysis results

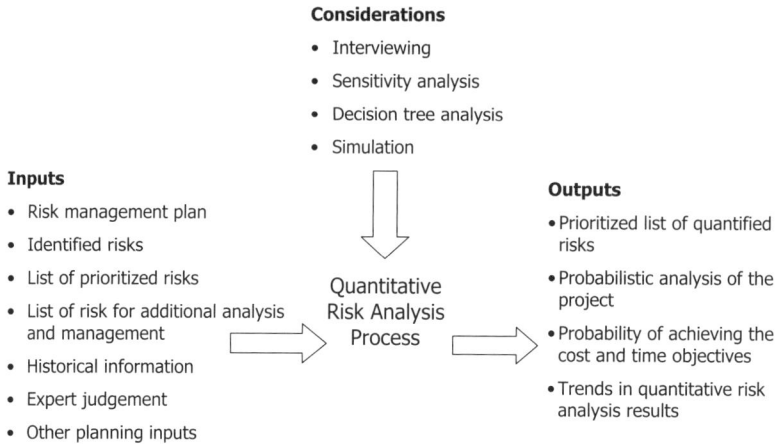

Figure 3.7   Quantitative risk analysis process

When following through the steps of the risk evaluation and management process, risks are first identified, and then evaluated in terms of consequence and probability. This means that subsequent effort, in terms of detailed assessment and demonstration of control, can be prioritized towards the most significant risks. One type of a detailed risk evaluation technique is known as the bow-tie analysis which, although is not restricted to risks with major consequences, it is most often applied when the risks lead to significant consequences.

The bow-tie methodology is a qualitative hazard assessment technique which encourages workforce involvement in analyzing the hazard scenario and provides a simple tool for straightforward communication of how the hazard is released, how it can escalate and finally how it is managed. The methodology is widely used in the oil and gas industry throughout the world where the implications of a risk occurrence may have serious or catastrophic consequences. The elements of a bow-tie diagram are shown in Figure 3.8.

The bow-tie diagram provides a 'snap shot' of the causes and consequences of the major hazard scenario and the controls in place at the time of the analysis to prevent the event or limit its impact. Once the assessment has been completed, the situation may change and, in particular, control measures may become less effective or disappear altogether. In order to provide assurance that the hazard will continue to be managed effectively, it is necessary to identify tasks which are carried out as part of the workforce's day-to-day duties which support and maintain the identified barriers, recovery measures and escalation factor controls. These critical tasks ensure that the control measures will continue to

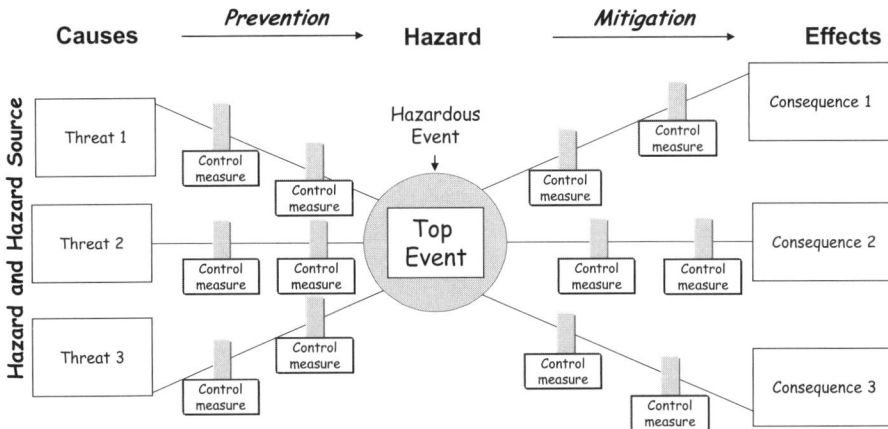

Figure 3.8   Elements of a bow-tie diagram

function in the future and are therefore essential to the ongoing management of the hazard. Critical tasks may be design activities, operations or maintenance activities or indeed those related to management and administration tasks. On this premise critical tasks are therefore not necessarily hazardous tasks.

Whilst there are no simple rules about the level of detail, the general tasks should be identified at such a level that they can be verified at a supervisor level. If they are targeted at too high a level (for example, if the managing director is made responsible for all tasks), this then becomes meaningless and also ineffective. Similarly, if tasks are assigned at too low a level, then the number of individual tasks to be documented becomes unmanageable.

In looking at the tools for qualitative analysis, one of the methods used is adapted and extended from the quantitative FMEA tabular tool and is known as Failure Mode Effect Criticality Analysis (FMECA). In this tool every item is ranked on a scale of 1 to 5 with the highest indicating the greater degree of significance. The entries may be those from the risk analysis or from a risk committee or brain-storming exercise. In the table (Figure 3.9) the term 'detection difficulty' refers to the perceived difficulty of noticing the cause of the risk, such as design error, in time to prevent the risk event from occurring. This requires a considerable amount of judgement. The product of these parameters gives a total ranking of the risk and when undertaken on every item in the risk list allows the list to be filtered in descending order so that the risks with the highest priority for management action appear at the top of the list.

Item	Failure mode	Cause of failure	Effect	Chance	Severity	Detection difficulty	Total ranking
1 Main building	Building collapses during installation of plant.	Errors in floor loading calculations.	Personal injuries. Project delays. Loss of reputation.	2	1	3	6
2 Main building	Building collapses during installation of plant.	Floor slabs incorrectly poured.	Personal injuries. Project delays. Loss of reputation.	1	5	3	15

Figure 3.9   Part of an FMECA matrix

When all of the project risks have been identified, assessed and ranked they can then be documented in a formal tabulated Risk Register which allows for review and communication of the risks with the project team, including the actions that are planned to mitigate the risk occurrence. In general a risk register is similar in format to the FMECA method shown in Figure 3.9. A risk register, as shown in Figure 3.1, can cover more detail than the table implies as it can discuss at length the nature of the risk, the impact and the actions which can be done to prevent or reduce the impact of the risk.

In looking at contingency planning for the project, contingency can be divided into two types: Technical and Managerial. Technical contingency covers the allocation of funds for specific risks which are outside of the normal project performance; on the other hand, management contingency covers the allocation of funds for risk areas that are not yet sufficiently understood to identify specific risks – for example new technology, build processes or new partnership arrangements. A contingency plan is applied to identify risks that arise during the project and developing such a plan in advance can greatly reduce the cost of an action should the risk occur. Risk triggers, such as missing an interim milestone for example, should warrant a fallback plan to be actioned. In general, a fallback plan is developed if the risk has a high impact on the project or on realizing its stakeholder's objectives. This approach might include an allocation of a contingency amount, development of alternative options or changing the scope of work.

The most usual risk acceptance response is to establish a contingency allowance (or reserves) including allocation of time, money or resources to account for known risks. The allowance should be determined by the impacts

Risk ID	Date	Risk description and consequences	Probability P = 1 - 3	Impact (severity) S = 1 - 3	Detection difficulty D = 1 - 3	Ranking P × S × D	Mitigating or avoiding action	Action by:

Figure 3.10   Format of a risk register

and computed at an acceptable level of risk exposure for those risks that have been accepted.

Cost contingency is established using a method such as Monte Carlo simulation (sometimes referred to as Monte Carlo Risk Analysis). This normally happens at the very early design point in the project life. Contingency is often established at an 80/20 confidence level, meaning that 80 per cent of the time the project can be completed at or below the funded level. By the same token it also means that 20 per cent of the time the project cannot be completed within budget. In practice, contingency should never be used to cover additional scope or changes to the agreed scope of work from the original project brief as this should be separately identified and funded as additional work. However, contingency may be used to cover the cost of performing activities that have been planed to assist the project in the management of, or mitigating the realization of, an identified risk.

As part of the project management activities, project managers should develop schedules with well-defined completion dates as 'cost' is often a function of time. All tasks with high-risk potential that will adversely impact the project schedule should have schedule contingency to fund additional time requirements. It is advised that schedule contingency should not be considered to be project schedule float and should not therefore be managed as float. Often this schedule float is used as a tool to manage schedule risk and *vice versa*. A critical path schedule driven purely by the logic of the work relationships and their associated durations will have float as its natural by-product.

Schedule contingency can be developed in four ways:

1.   **Monte Carlo simulation** – This method employs the same type of Monte Carlo simulation used in developing cost contingency to determine schedule contingency. For each milestone a target

value, a high estimate, a low estimate and a probability factor is assigned. The probability factor is expressed as a percentage and is the probability that the actual outcome will yield a range of results from which the project manager can choose the desired probability of success.

2. **Monte Carlo simulation of the critical path** – This method is conducted in the same way as described above but uses all of the activities on the critical path. A full project critical path schedule is needed for this method and is often not achievable during the concept phase of the project when the schedule contingency is still being developed.

3. **Project Evaluation and Review Technique (PERT)** – PERT is a widely used scheduling method that employs use of three durations for each schedule activity. These three durations represent the most pessimistic, the most probable and the most optimistic time elements. In order to determine the expected duration of the project, the following formula is then used to provide a weighted expected duration:

$$\text{Duration} = \frac{\text{most pessimistic} + \text{most optimistic} + 4\,(\text{most probable})}{6}$$

4. **Historical/past experience** – This method employs the use of the experience and expertise of the members of the project team. In this method the project duration is based on their experience, education and judgement from work on similar projects. If this method is not supported by site-specific historical data then it should not be used because its output results are largely subjective and justifying the amount of contingency may prove to be difficult.

As noted above in the PERT method, the use of three point estimates is a common approach in looking at schedule risk. For each project the following questions will need to be answered:

- *On the most probable or expected date* – what the assumptions are, and what interdependencies must be realized to ensure that the required date will be achieved?

- *For the best case (sometimes referred to as the 'target' date)* – what opportunities exist which can be used to shorten the activity and allow the milestone date to be achieved earlier, and by how much?

- *When looking at the worst case date* – what are the schedule-related risks that, if left unmitigated, would delay the activity and the achievement of the milestone event, and by how much?

Three point estimating can also make use of the following formula to allow an expected value to be calculated using the same formula as shown above. For example:

$$\text{Expected value} = \frac{\text{most optimistic (MO)} + 4\,(\text{most likely (ML)}) + \text{most pessimistic (MP)}}{6}$$

or another weighted version is that the:

$$\text{Expected value} = \frac{\text{MO} + 3(\text{ML}) + 2\,(\text{MP})}{6}$$

This also leads into understanding the degree of uncertainty where (MP-MO)/6 = the standard deviation, and the topic of statistics and the normal distribution rule.

Finally, sensitivity analysis helps to determine which risks have the most potential impact on the project by examining the extent to which the uncertainty of each project task affects the project from being achieved. In undertaking quantitative evaluation, all project models are comprised of variables and formula relationships where some can be weighted to be more important than others to determine how the project can be modelled and provide the best guide to management planning and decision taking.

Risk Response Planning, Monitoring and Control

Risk Response Planning

Having undertaken the risk analysis and evaluation described in the previous chapter this leads the project or risk manager to now look at risk mitigation by taking appropriate actions to achieve the project's objectives through revision to the project's schedule, budget, scope or quality. Risk management should therefore be regarded as an integral part of project management and not as an additional extra. These final phases of risk management involve establishing specific action plans to manage the risks and, more importantly, the identification of fall-back plans to drive, inform and support risk response planning.

Effective risk management demands an active process of regular risk reviews and the commitment to:

- anticipate and influence events before they happen by taking a proactive approach;

- provide knowledge and information about predicted events;

- inform and, where possible, improve the quality of decision making;

- avoid covert assumptions and false definition of risks;

- make the project management process clear and transparent;

- assist in the delivery of project objectives in terms of benchmarked quality, time and cost thresholds;

- allow the development of scenario planning in the event of the identification of a high-impact risk;

- provide improved contingency planning;

- provide verifiable records of risk planning and risk control.

To achieve risk management which is not only effective but efficient requires risk response planning. The commonest form of risk response planning is the Risk action plan where the inputs to this include:

- the risk management plan;

- a list of prioritized risks;

- a risk ranking of the project;

- a prioritized list of quantified risks;

- a probability analysis of the project;

- a list of potential responses as the risk identification process can help suggest a response to individual risks or categories of risks;

- the level of risk that the project stakeholders are able to own. As risk owners they should be involved in developing the risk response;

- any trends from the qualitative and quantitative risk analysis results. Trends in the results can make the risk response, or further analysis, more or less urgent and important.

There are a number of tools and techniques which can be employed in the risk response planning phase and as discussed in Chapter 1 the options available as actions to risk are based on one or more of the '4Ts' risk response actions, namely to Terminate, Treat, Tolerate and Transfer risks (see Figure 4.1).

Where it is not possible to reduce the risk probability, a mitigation response (that is, 'treat' on the 4T's) might address the risk impact by targeting linkages that determine the severity. The outputs of risk response planning include the following:

Identification of residual risks	Residual risks are those that remain after the 4T's actions have been taken. These may also include minor risks that have been accepted and addressed, for example, by adding contingency amounts to the costs or the project time allowed.
Identification of secondary risks	Risks that arise as a direct result of implementing a 4T risk response are termed as secondary risks. Having identified these then suitable responses need to be planned and managed.
Detail contractual agreements	Contractual agreements may be entered into to specify each party's responsibility for specific risks should they occur. These will also include insurance and other items as appropriate to avoid or mitigate threats.

Identify contingency	Contingency reserve amounts that may be needed will need to be allocated. The probabilistic analysis of the project and the risk thresholds help the project manager to determine the amount of contingency needed to reduce the risk of overruns of the project objectives to a level acceptable to the organization.
Detail inputs to other processes	Most responses to risk involve expenditure of additional time, cost or resources and require a level of assurance that spending is justified for the level of risk reduction. Alternative strategies must be fed back into the appropriate processes in other knowledgeable areas.
Detail inputs to a revised project plan	The results of the response planning process must be incorporated into the project plan to ensure that agreed actions are implemented and monitored as part of the overall project.

Several risk response strategies are often available for consideration by the project team and the strategy that is most likely to be effective should be selected for each risk. Specific actions can then be developed to implement each strategy. However, it is advisable to have both primary and back-up secondary strategies selected.

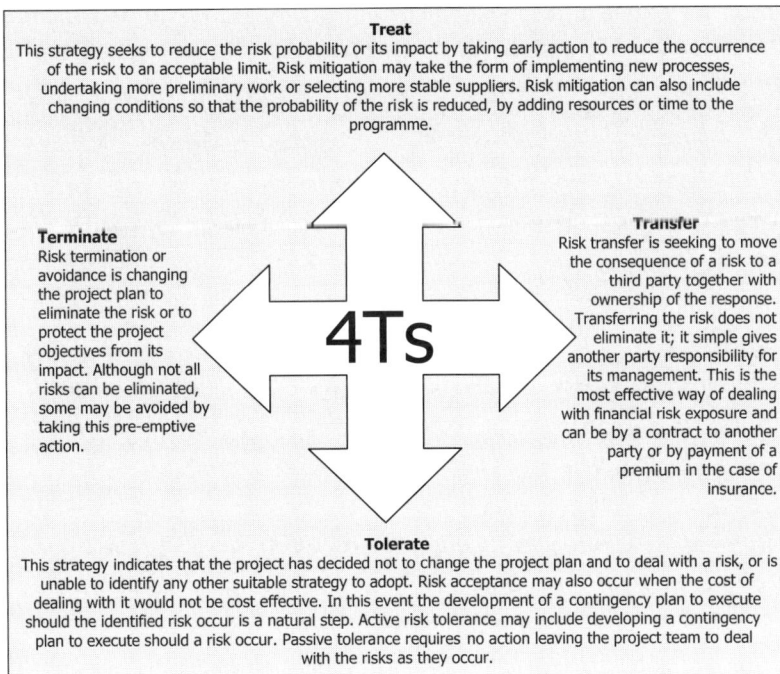

Treat
This strategy seeks to reduce the risk probability or its impact by taking early action to reduce the occurrence of the risk to an acceptable limit. Risk mitigation may take the form of implementing new processes, undertaking more preliminary work or selecting more stable suppliers. Risk mitigation can also include changing conditions so that the probability of the risk is reduced, by adding resources or time to the programme.

Terminate
Risk termination or avoidance is changing the project plan to eliminate the risk or to protect the project objectives from its impact. Although not all risks can be eliminated, some may be avoided by taking this pre-emptive action.

4Ts

Transfer
Risk transfer is seeking to move the consequence of a risk to a third party together with ownership of the response. Transferring the risk does not eliminate it; it simple gives another party responsibility for its management. This is the most effective way of dealing with financial risk exposure and can be by a contract to another party or by payment of a premium in the case of insurance.

Tolerate
This strategy indicates that the project has decided not to change the project plan and to deal with a risk, or is unable to identify any other suitable strategy to adopt. Risk acceptance may also occur when the cost of dealing with it would not be cost effective. In this event the development of a contingency plan to execute should the identified risk occur is a natural step. Active risk tolerance may include developing a contingency plan to execute should a risk occur. Passive tolerance requires no action leaving the project team to deal with the risks as they occur.

Figure 4.1   Risk response actions

Source: Figure developed from 'The Essential Management Toolbox', S.A. Burtonshaw-Gunn (2008). © John Wiley and Sons. Reproduced with permission.

Action plans are needed for all significant risks identified and in particular those with a high criticality from the impact matrices. The action plans must be cost effective and are likely to use the technique of three point estimates as covered at the end of Chapter 3. The contents of a typical action plan cover:

- description of the planned activities;

- ownership;

- start date and any identified risk trigger conditions;

- the cost and resource requirements;

- secondary risks;

- comments and the action status, employing the widely used 'BRAG' coding (where blue means completed action, green means on plan, amber indicates that it may not be achieved and red that it is unlikely to be achieved).

For whatever reason, if the mitigation action is not proving successful and the risks become a problem then a Fallback Plan will need to be initiated. Although some consider this an unnecessary luxury there is a great benefit to having fallback plans already developed with time to then quickly put them into service. The contents of a fallback plan are similar to the mitigation action plan, namely:

- description of the plan for recovery and clear ownership;

- start date/trigger condition;

- the costs and resource requirements;

- secondary risks as a consequence of the fallback plan's actions;

- comments and action status, using the BRAG coding when implemented.

It has to be recognized that on some occasions there can be no effective, or cost-effective, fallback recovery plan. Residual risks may also be present and rarely is it cost-effective to seek to remove risks entirely; therefore even after action plans have been initiated some small level of risk is likely to remain. It is common that action plans or fallback plans introduce risks of their own; these secondary risks also need to be documented and accounted for in assessing the cost-benefit of their response.

Despite all this risk identification and planning of actions, no industrial activity is entirely free from risk and so many organizations and regulators around the world require that safety risks are reduced to levels that are 'As Low As Reasonably Practicable', (ALARP). The 'ALARP region' lies between unacceptably high and negligible risk levels. Even if a level of risk has been judged to be in this ALARP region it is still necessary to consider introducing further risk reduction measures to address the remaining, or residual, risk. The ALARP level is reached when the time, effort and the cost of further risk reduction measures become unreasonably disproportionate to the additional risk reduction obtained.

Project risks can be reduced by employing the 4T strategies already discussed or increasing the number and effectiveness of controls. The concept stage of a new project offers the greatest opportunity to achieve the lowest residual risk by considering alternative options. This is a standard approach in offshore oilfield development and once the concept is selected and the design progresses, the attention moves to considering alternative layout and system options to maintain risk reduction and optimize safety. In this industry example, the attention of the safety management culture is on collecting feedback, improving procedures and managing change to maintain the residual risk at its ALARP level. However, with advances in technology, what is accepted as ALARP at the design stage may not be regarded as such when the plant becomes operational, warranting necessary periodic risk reviews.

Working to the ALARP principle means that risks are reduced to a level that is 'as low as reasonably practicable' and which can be considered to have been achieved when:

- legislation, established standards and good design practice have been complied with; and

- that the resources (in terms of cost, time, difficulty and risk) required for the implementation of additional measures which may further reduce the risk are disproportionately large when compared to the potential benefit to be gained. This will naturally suggest when resources would be better applied to then reduce risks in another area of the business.

For those high-risk industries where ALARP is a major feature of the risk management process the first stage will be to demonstrate that the risks are within established acceptance criteria for risk tolerability (that is, they are not inside the intolerable zone shown in Figure 4.2)

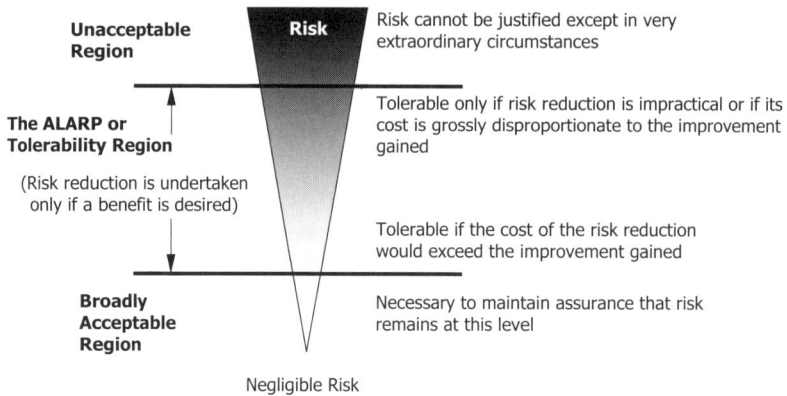

Figure 4.2   ALARP risk zones

These criteria may be legislative, industry or company standards and the risks can be demonstrated to be tolerable by reference to relevant industry best practice, professional and, where necessary, suitable quantitative assessment. Once it has been determined that the risks are in the tolerable region, the next stage of the ALARP demonstration will be to determine if there are any additional, practicable risk reduction measures which are reasonable to implement. All risk reduction measures will be assessed to determine whether they are technically viable and offer a significant benefit.

In many situations, such assessments can be based simply on professional judgement, experience and recognized best practice. In other situations, the effort required to implement a risk reducing measure, in terms of cost, time, difficulty, resources required and so on, needs to be formally evaluated against the risk benefit likely to be achieved. Where the effort is shown to be grossly disproportionate to the benefit, the measure can then be rejected. Where the effort is not disproportionate, the measure should be implemented immediately or at the most convenient point in time especially where the risk impacts of health and safety.

During design development, demonstration of ALARP is inherent throughout the process and relies on the assessment and conclusions reached by the project team. By asking the questions: 'Are the proposed control arrangements good enough?' and 'Can we do anything better?' the team uses its experience and expertise to verify that the residual risk associated with the design is managed to ALARP levels.

Moving to the operations stage, ALARP can be achieved by reviewing the existing control measures and questioning if anymore can be done. In this case, improvements may involve redesign or modification, but also changes to procedures, processes and even established ways of working.

Figure 4.3 illustrates the relationship between risk and the resources required to reduce its occurrence and how the ALARP position segregates the costs of mitigation from the residual hazard.

ALARP may be assessed in several ways, varying from documenting the purely subjective evaluation based on engineering judgement to taking a fully quantitative approach where risk reduction is expressed as a measurable parameter and formally evaluated against cost.

Risk Monitoring and Control during the Project

During the project, risk monitoring and control is the processes of keeping track of the identified risks, monitoring the residual risks and identifying new risks. This process should also ensure the execution of the risk plan and continually evaluate the plan's effectiveness in reducing risk. Resource allocations can also be monitored as these too will have been pre-planned and, where appropriate, allocated to the agreed actions. Immediate risk actions should be built in with the other project activities as an integral part of the overall project management plan. Other actions will be dependent upon the risk materializing and will be triggered by the occurrence of risk metrics and milestone events.

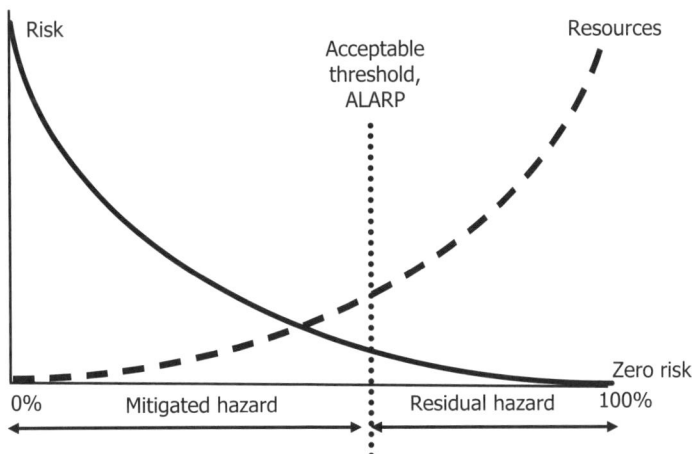

Figure 4.3   ALARP principles

Risk monitoring and control also records risk metrics associated with implementing contingency plans and needs to be, of course, an ongoing process for the duration of the project. Naturally the risks change as the project matures; new risks may develop or the anticipated risks disappear.

The elements of the process of risk monitoring and control are shown in Figure 4.4.

Attention to the risk monitoring and control processes will provide information that can assist with making effective decisions in advance of the risks occurring. Communication with all project stakeholders is needed to periodically assess the acceptability of the level of risk on the project. The purpose of risk monitoring is to determine:

- that risk responses have been implemented as planned;

- that the risk response actions are as effective as expected, or if new responses should be developed;

- that the project assumptions are still valid;

- if the risk exposure has changed from its prior state with additional analysis or trends;

- if a risk trigger has occurred;

- if the correct policies and procedures are followed;

- if any previously identified risks have occurred.

Considerations

- Project risk response audits
- Periodic risk reviews
- Earned value analysis
- Critical design features management
- Design reviews
- Additional risk response planning

Inputs

- Risk management plan
- Risk action plan
- Project communication
- Additional risk identification and analysis
- Scope changes

Risk Monitoring and Control

Outputs

- Fallback plan
- Corrective actions
- Project change requests
- Updates to the risk action plan
- Risk database
- Updates to risk identification checklists

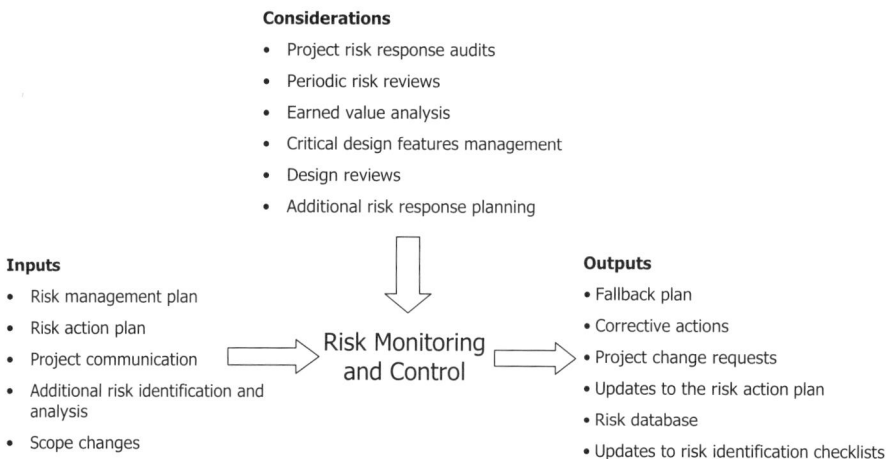

Figure 4.4   Elements of the risk monitoring and control process

Risk control may involve choosing alternative strategies, implementing a contingency plan, taking corrective action(s), or even the re-planning the project. It is important for the risk response owner to report periodically to both the project manager and the risk team leader on the effectiveness of the plan, any unanticipated unwanted effects and any mid-project programme correction needed to mitigate the risk occurrence.

Some activities of risk monitoring and control use triggers to indicate that a risk is occurring as defined during the risk response planning stage. This is one of the outputs of the action plan and as such triggers are events or consequences that cause activation of the corrective actions: a fallback plan, updating of the risk plan and use of the risk identification checklists. The concept of monitoring and feedback of risk information is a standard systems approach as shown in Figure 4.5.

The final stage of risk control is to:

- monitor risk metrics and milestones so that, if required, contingent actions can be implemented;

- monitor the effectiveness of risk management actions to ensure that they are having the predicted effect;

- feedback lessons about which actions are the most effective.

There are a number of project control tools which are very effective as indicators in managing project risks. In looking at the closed loop feedback system (Figure 4.5) it is noted that this is applicable in a range of project control areas.

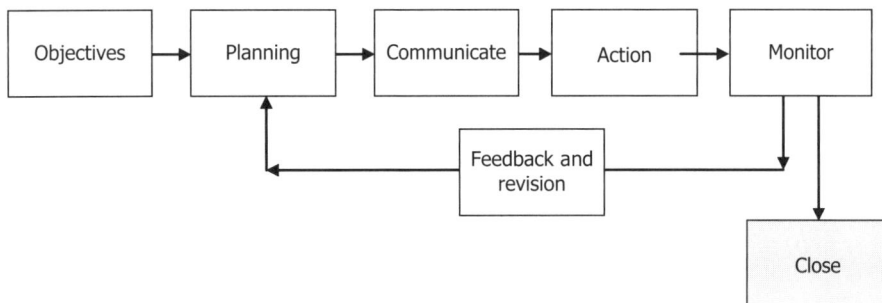

Figure 4.5   Risk monitoring and control: the use of a closed loop feedback system

Source: From '*The Essential Management Toolbox*', S.A. Burtonshaw-Gunn (2008). © John Wiley and Sons. Reproduced with permission.

When choosing the object of this control loop application, the project manager must consider the value of the control provided and judge this against the cost of obtaining it. The factors that need to be considered are:

- the degree of detail needed to provide the desired status or performance information;

- the frequency of the feedback needed about the project status or its performance;

- the accuracy of the feedback of the measurement required to provide the desired status or performance information;

- the timeliness of the feedback including how soon the data is required to support decision making on the project;

- the level of management attention required to obtain the information, the type of records that must be maintained and the format in which information must be provided;

- the cost of producing and using the data. Generally the greater the frequency, the level of detail, accuracy and timeliness all serve to increase the cost of the process.

As new risks are identified throughout the project it is possible that the increased project exposure will exceed its available contingency. However, risks can be 'retired' when they occur and these become business issues to be addressed by the fallback plan or when the risk window date has passed without the risk occurring due to either good management, a change of circumstances or simply the passage of time. On the basis of the retired risks, the contingency set aside for them should be released at that point and made available as contingency for any new risk which may immerge. On this basis for the risk management plan to be of maximum benefit, the project must be monitored against it and this should typically involve:

- coordinating the risk activities alongside the other project work;

- monitoring resource usage against limits and resolving any conflicts;

- monitoring risks to ensure that they remain within the agreed project limits;

- monitoring risks to ensure that they do not become too large to manage effectively;

- monitoring risks to ensure that they do not threaten the viability of the project objectives;

- re-evaluating the risks to the project as outlined in the risk strategy;

- implementing an agreed problem resolution process for any risks or issues that fall outside the authority of the project team. For example, an environmental risk which may impact on more than one of the project's stakeholders;

- implementing an agreed problem resolution process for any risks or issues that are not being effectively managed by the risk owner, and as a consequence, other stakeholders who may be adversely affected. Such problem-solving escalation demonstrates good project management; it should not be viewed as an admission of failure.

Risk Monitoring and Control at Project Closure

Having looked at risk monitoring and control during the project it has to be said that project closure is also an integral part of the construction process lifecycle; although in practice its importance is often underestimated, or even ignored. Formal closing of a construction project warrants its completion in an organized and controlled manner. This phase usually includes the following four activities:

1. obtaining acceptance of the project results from the client;

2. balancing the project budget;

3. closing of the final account and invoice payments, warranties and as-built details;

4. conducting of a project closure meeting.

Importantly, project closure also presents the opportunity to learn from the project experience for the benefit of both the individuals and the companies involved with respect to any future similar projects. As such a fifth activity is suggested as:

5. documenting the experience gathered.

As part of the post-project review process, a questionnaire can be distributed among all project participants (for instance, the members of the design and supervision teams, client representatives, suppliers and other project participants) aimed at gaining a number of individual assessments of the specific aspects of design planning and implementation of the project. The information obtained using questionnaires can be compiled and summarized in a final project report and then distributed among the members of the project team and discussed during the project closure meeting. A report from the post-project study can facilitate discussion on the project experience – what has gone right, what went contrary to the plan and what should be improved. This may also be known as Learning from Experience (LfE), and should be documented and stored in an accessible repository of the organization's project knowledge. It may then serve as a basis for assessment of processes, implementation of improvements and modified cost and risk allocation contingencies for other potential projects.

Project closure, where used, may follow company or industry processes such as:

a)   The set of 'good practices', for example the document issued by PMI (Project Management Institute) or the Association of Project Management's *Body of Knowledge*.

b)   The PRINCE 2 Project Management methodology.

Both of these processes are discussed below.

Project closure in accordance with the APMs *Body of Knowledge* consists of two main processes:

1.   Closing the project. This is a process necessary to finalize the activities in all groups of project management processes. It allows the project or a project stage to be formally closed and includes an opportunity for the completion acceptance of all works and the gathering of project documentation. Subject to assessment are also the problems of failure on the project (or its stage), an opportunity to review the areas of success on the project and produce some conclusions applicable to further work.

2.   Contract closure. This is a process necessary to finalize the contractual requirements covering the provision of goods or professional services pertaining to the whole project or a particular contraction phase or activity. This procedure is associated with the

actual checking of the project result as well as financial settlement of the project (or its stage).

Project closure in accordance with the PRINCE 2 methodology calls for an assessment of the project effectiveness (and its profitability) and uses the 'business case' term. Specification of the business case is to answer how the project assists the development of the company and questions if the expected results are worth the dedicated time and money invested – not just the profit generated. The document describing the original business objective will serve as a basis for comparison with the actual situation, and thus aid the final financial settlement of the project. The methodology establishes, among other things, the following three key criteria necessary for closing projects:

1. Has the project been constructed as required and have all of the products specified in the project plan and the WBS been delivered, installed and shown to be functional?

2. Are the project resources and support finished functions now and no longer needed?

3. Are there any contractual consequences on closure of the project?

Answers to the above three questions will be contained in the project completion report prepared by the project manager and aimed at summarizing the project's performance by comparison with the documents prepared earlier; in particular these documents will include the project plan and the risk plan as discussed in Chapter 2, together with the cost plan and project programme.

Whilst the above suggestions, in the main, refer to the risk management at project end, the opportunity also exists to review the lessons, knowledge and experiential learning about the effectiveness of the project's wider actions. For example, a preventative action may have been put in place to reduce the likelihood of a risk from high to medium and the project closure review presents an ideal opportunity to question:

- if this was effective?

- did the risks materialize?

- if so, was this because the action was ineffective or for some other reason?

- in retrospect, might a different action have been more effective?

Such lessons can be used to:

- improve ongoing risk planning for the organization's future projects;

- share good practice with other current projects: for example through project archives, project evaluation reviews or post implementation reviews;

- monitor risk metrics and milestones.

Project closure will therefore involve both checking whether risk metrics or milestones have been reached, whether any contingent management actions are required, and ensuring that indicators continue to give a clear picture of the status of project risks until those too are closed off. The roles and responsibilities in the post-project process are presented in Figure 4.6.

Role	Obligations	Responsibility	Rights
Project Management Office	Prepares, analyzes, interprets and distributes results of post-project study in form of a report. Updates the project knowledge database by adding new, unique knowledge.	Organizes the post-project review and present the report on experience gathered.	Present an objective assessment of the project on the basis of the post-project study results.
Project Team Members	Fill out post-project questionnaires and participate in meetings.	Fill out the post-project questionnaires on time, providing honest answers. Participate in discussion on experience gathered.	Formulate process improvement recommendations.
Supervising Team Members	Fill out post-project questionnaires and participate in meetings.	Fill out the post-project questionnaires on time, providing honest answers. Participate in discussion on experience gathered.	Formulate process improvement recommendations.
Project Parties	Fill out post-project questionnaires and participate in meetings.	Fill out the post-project questionnaires on time, providing honest answers. Participate in discussion on experience gathered.	Formulate process improvement recommendations.
Suppliers	Fill out post-project questionnaires.	Fill out the post-project questionnaires on time, providing honest answers.	Formulate, from the perspective of the suppliers, their own process improvement recommendations.

Figure 4.6   Responsibility matrix: post-project review process

Construction Prime Contracting and the Importance of Risk Management in International Projects

CHAPTER

5

The Growth of Prime Contracting

In parallel with an increase in collaborative working such as 'partnering' between the main contractor and the client organization and closer supply chain management since the Latham and Egan reports in the 1990s has been a growth in Prime Contracting based upon supply chain integration. This initiative has also been assisted by the development of the prime contracting approach which has been recommended by the United Kingdom Treasury and the National Audit Office for the procurement of central government construction projects through the Private Finance Initiative (PFI) and Public Private Partnership (PPP). Both these project finance routes bring together the infrastructure project management of design and construction with external private sector funding arrangements. On these projects such financing is typically through bank loans, equity provision/exchange or investment through a number of innovative funding arrangements when traditional client funding is unavailable or when other priorities have stronger demands on a government department's limited finances. Under such PFI projects, the private sector involvement varies but in the main builds, finances and operates the public infrastructure such as roads, rail links and airports recovering the cost through service provision charges (sometimes referred to as 'power by the hour'). Whilst there are a number of variations of these types of projects, each necessitates a long-term commitment between the facility provider, the user and the ultimate client who only acquires ownership of the facility at the expiry of a significant contracted period when the revenues have repaid the original project costs, the operational costs and associated profits. Whilst successful management of construction projects presents a challenge in any environment, the topic of this chapter is to present an understanding of the importance of risk management in such PFI infrastructure projects and then to examine those specific risks appropriate to such projects in an international setting.

The growth in collaborative partnering and for some companies an evolution into prime contracting have arisen from two principle government-supported industry reviews reported in the publications 'Constructing the Team' by Sir Michael Latham and 'Rethinking Construction' by Sir John Egan, published in 1994 and 1998 respectively. These two publications have supported an almost universal client requirement in both the private and public sectors to achieve the benefits of increased value in infrastructure projects from both a facilities management and an ergonomic perspective. As a first step many clients have moved away from competitive tendering to favour the use of contracts with closer supply chain management through advances in both project and long-term strategic partnering arrangements between clients and their facilities providers, constructors, designers and, in a few cases, with a number of their second-tier suppliers. Establishing close and strong supply chain relationships is the ideal for any company, however this is not within risk and the risks of Supply Chain Management and collaborative working can be seen in Figure 5.1. Having mentioned the increase in 'partnering' the actual extent of this commitment has been found to vary in practice with the largest proportion of such arrangements being confined to the main construction companies, and their first tier suppliers.

It should be noted that within the construction industry there is no uniform definition of the term prime contracting which appears open to interpretation by constructors, clients and project financiers, with each offering a view which often supports and promotes their own roles. From the government point of view the National Audit office defines prime contracting as:

> 'A contract involving a main-supplier, the Prime Contractor, which has a well established supply chain of reliable suppliers of quality products to encourage increased quality and value for money resulting from an element of consistency and standardisation.'

> Source: National Audit Office (2003) PFI: Construction Performance. Report by the Comptoller and Auditor General, The Stationery Office, London.

One of its main attributes of this approach is seen in the very term prime contracting which endorses the focus on a single point of responsibility for the design, building, operation and, sometimes, maintenance of the facilities until making the ultimate delivery of the project to the client. Although there is no universal form of a Prime Contract *per se* those contracts that are used

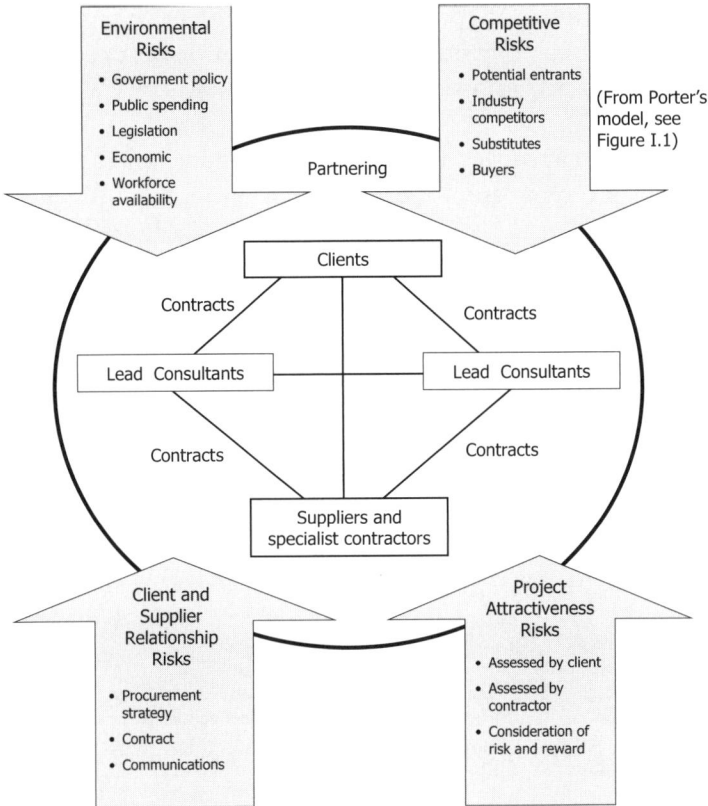

Figure 5.1   Risk and supply chain management

Source: From S.A. Burtonshaw-Gunn (2004) 'Examining risk and supply chain collaborative working in the UK construction industry' in *Supply Chain Risk*, Editor Professor C. S. Brindley, Published with permission of Ashgate Publishing.

to commercially formalize the arrangement have a number of similarities in their intent to provide opportunities for improved efficiencies and savings throughout the contract period. These savings are typically achieved through coordination and integration of the activities of a number of subcontractors aimed to meet the overall specification, on-time delivery and efficient operation of the facility.

With an increasing reliance in the use of private sector funded development and long-term operation of major projects around the world, such as PPP projects, the single point of accountability – the Prime Contractor – now needs to consider the through-life costs to a far greater extent than when merely providing a new facility for a client without the added responsibility of its day-to-day operation and longer-term maintenance costs. As such, just like partnering, the adoption of prime contracting, rather than past traditional procurement based on competitively tendered and lowest cost, represents a major advance in terms of improving value for money.

This contractual approach will typically encourage an early involvement of the total supply chain from facility design through to its operation as a way to improve performance and reduce project risk. Such a move to early involvement of the suppliers also provides advantages in planning and risk reduction of delays in acquiring long lead time items. The areas of management which prime contracting is likely to cover are shown in Figure 5.2. Here the focus is on the financial considerations for privately funded projects, such as those undertaken under the PFI and PPP procurement routes, and is shown as a function of commercial management.

In addition Figure 5.2 also shows the relationship between the attributes of a single point responsibility of the prime contractor and risk management

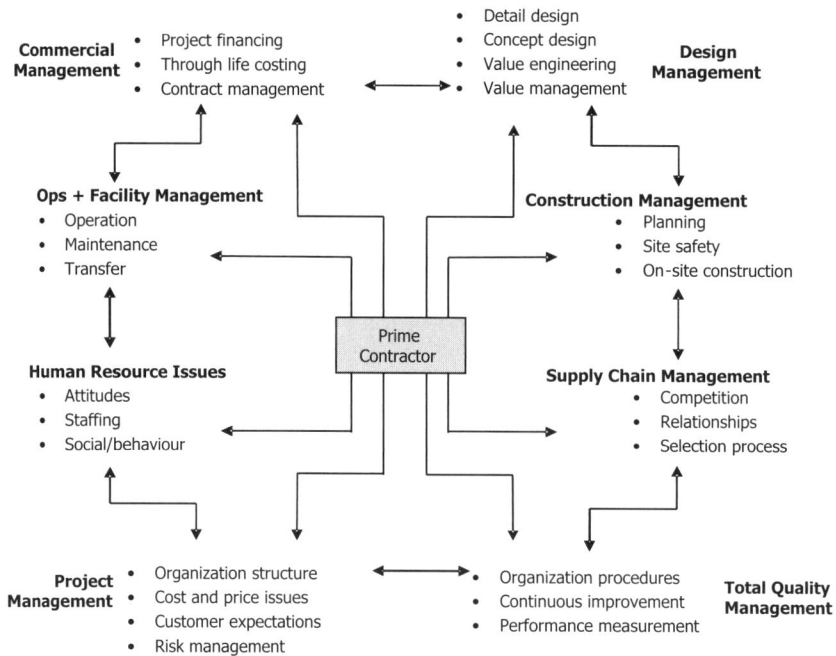

Figure 5.2   Construction prime contracting and its relationship with other management disciplines

Source: Considerations of Pre-Contract Risks in International PFI Projects, Professor S.A. Burtonshaw-Gunn (2005). Reproduced with kind permission of the Salford Centre for Research and Innovation, University of Salford, UK.

considerations. This is shown in the figure as a sub-set of Project Management although it may also reside as part of the Commercial Management activities. In executing a typical development and operation project arrangement through a prime contracting procurement method, one of the key advantages is the opportunity for the risks associated with the project to be devolved to the party best able to deal with them: this delegated authority could be the client or the contractor, or indeed an appropriate member of the project supply chain.

Whilst it is widely accepted within the construction industry that the number of client-customer interfaces and the imprecise allocation of risks have long since hampered traditional procurement, the use of a prime contracting framework provides a degree of confidence with respect to project uncertainty together with a reduction in the amount of required contingency provision typically witnessed with the more traditional procurement methods. The use of prime contracting encourages and cultivates the development of non-adversarial collaborative working also witnessed in Construction Partnering; this can foster a more systematic approach to Value Engineering, Value Management and Risk Management by employing multi-party workshops as a key activity in assessing the options to achieve the required fitness for purpose quality standard and hence reduce construction, commercial and operational risks.

The increase in magnitude in both financial scale and often complexity of prime contracting projects requires clear identification of the project risks throughout the supply chain and is benefited by a common approach to the process of allocating risks to where they can best be addressed. This is undertaken against the premise that risk management is undertaken to ensure that the project is executed in accordance with a plan which in turn allows its agreed objectives to be achieved. Although the way risk management is operationalized varies from one company to another, the most common approach is a staged approval of risk identification followed by employment of a number of strategies for its management. Whilst covered in the previous chapters, this approach covers risk identification to capture the potential risks which could arise within the project followed by risk classification where risks are grouped into internal risks and external risks which are outside its direct control. The next step is to undertake a risk analysis to quantify and evaluate the risk on the project leading to the final stage of risk response which addresses how the risk will be managed. This final activity could cover a range of actions including risk transfer, treatment, termination or toleration as discussed in Chapter 4 and shown in Figure 4.1 of that chapter. To assist in this project management task there are a number of proprietary project planning tools which can be supplemented by add-on or stand-alone risk management programmes providing risk simulation, scenario and sensitivity analysis covering the individual project stages with accumulation over the whole project.

Consideration and Identification of Risks on International Projects

Having examined the role of the prime contractor and the topic of construction risk management in a general way, the focus of this chapter now moves to examine the risks associated with undertaking major project work in an international environment. Whether in mature, developing or underdeveloped regions, one common requirement is the necessity for countries to construct, repair, refurbish and modernize their infrastructure. An increasing number of governments continue to consider using prime contracting in conjunction with private funded project development and operation. Indeed, as international interest has grown in this strategy a number of different forms of PFI arrangements to accommodate Foreign Direct Investment (FDI), long-term leasing and private funded ownership of public infrastructure facilities have evolved. In addition, many countries have enacted legislation to facilitate the encouragement of FDI. The differences in the investment arrangements mostly centre on the timing of when ownership is transferred back to the public sector corporation or government department. The projects which are most often seen as suitable for private funding and operation before transfer typically range from roads, bridges, water systems, airports, ports and public buildings such as museums, prisons, hospitals and schools. Whilst this approach has increased in popularity over the last decade, the first example of a PFI project was used as far back as 1834 with the construction of the Suez Canal which was financed by European investment with Egyptian financial support for the design, construction and its revenue-producing operation.

The use of different PFI arrangements can be used in conjunction with prime contracting to allow private developers to design, finance, construct and operate revenue-producing public projects, and then return them to the state at the end of an agreed payback period. This period is always fixed as part of the contractual arrangement and usually ranges between 25 to 40 years. From a government point of view one of the reasons that such PFI schemes have become so popular is because the private sector builds the project with little or public investment and then at the end of the contract or concession period the public sector then obtains the facility with it becoming a state asset such as a road, railway or a more capital intensive project such as an airport. This ownership transfer occurs without the burden of any government funding.

Whilst construction project risks can typically be divided into three phases: pre-construction risk and planning, construction and commissioning risks, and finally, operational risks. (see also Chapter 6 with respect to the investment at these phases), it is the general, less technical but more important risks that are suggested to require careful consideration and management if an international PFI project is to have any chance of success. Clearly well before any potential prime contractor formally submits an expression of interest, holds discussions about a project – let alone a fully costed proposal – an outline understanding of indigenous political and economic landscape can serve to either attract or dampen the project's attractiveness. This understanding is important as it provides an initial 'snapshot' of the risks of undertaking general business in the identified international environment. These can be further developed by a PESTLE (Political, Economic, Social, Technological, Legal and Environmental) analysis to examine specific risks associated with political stability and any potential change; the laws and regulations associated with commercial activities in the project location together with any special technical standards and environmental issues which may adversely impact on the project to be undertaken. A further early stage investigation should be undertaken to examine national employment legislation which may affect the project programme and any inward investment requirements associated with the use of local and overseas employees, both of which may constrain the prime contractor's supply chain if local suppliers are expected or contractually required to be significantly involved as part of an employment provision.

Attracting the necessary project funding either through banks, government, or private investors for PFI projects, understanding the forecast revenue opportunities, charge basis and profit potential all need to be fully explored at the earliest stage to allow an understanding of any project constraints and to identify any risks to revenue generation. This understanding is necessary to reduce the occurrence of conflict of interest problems which typically affect project participants. In practice these are often seen in the differing demands of project facility operators and the project's financiers where the latter will be looking for the earliest possible financial return on their investment. Whilst reviewing the willingness and productiveness of the political structure, the suitability of prime contracting will need to be addressed, in particular the support for adopting a prime contracting approach with a single point of contact for the whole of the supply chain, effective management of a supply chain comprising local and foreign stakeholders and the management of the interfaces of the parties during the design, construction, commissioning and operational phases of the project. Given that prime contracting is seen to be appropriate, there will also need to be a high level of confidence by the host public or private client with respect to the acceptance of both formal and informal supply chain relationships with a number of raw material suppliers, main subcontractors, specialist service providers and, to a lesser extent, their own supply chains.

Thus it is suggested that the pre-construction risk analysis phase is clearly important to provide the most current information on which to base any strategic decisions of undertaking the potential work proposed. Whilst there will be the financial costs to the business in conducting this investigation it provides an informed basis on which it may base any future decision as to whether to prosecute the opportunity, based importantly not just on its expected fees and profit in isolation but the inherent risks involved. Another benefit of such a due diligence investigation is the use in assisting the process of raising project finance from banks or other lenders. It has to be recognized however that time and effort spent on such pre-bid investigations are no guarantee of securing a successful appointment for such projects.

For the successful bidder, the length of the concession period must be sufficient to allow the recovery of the investments for all of the funding parties; this is determined by the contractual agreement with the client. If the economy inhibits investment from private companies the project simply cannot be developed almost irrespective of any financial rewards. Indeed risk and reward have to be balanced and funding parties will need to be convinced that there is a need for the project/facility on which the business will generate a return on the investor's contributors. Only after a full market analysis that is able to justify a need for the facility will private investors be willing to consider providing finance and their participation in the project. The financial issues seen as important to shareholders – private investors and perhaps the government – are also regarded as crucial to lenders in order that they have sufficient confidence to be able to forecast the financial outcome. The areas of risk in international projects are shown in Figure 5.3.

On successful selection of a prime contractor and with the signing of the concession contract, the detailed risk management process can formally commence.

International projects often result in the need for a complex contractual finance mechanism with the establishment of an arrangement, often referred to as a Special Purpose Vehicle (SPV) to allow the operating contracts with key participants to be agreed. One of the key features of an SPV is the risk allocation and apportionment between the client and the prime contractor, subcontractors,

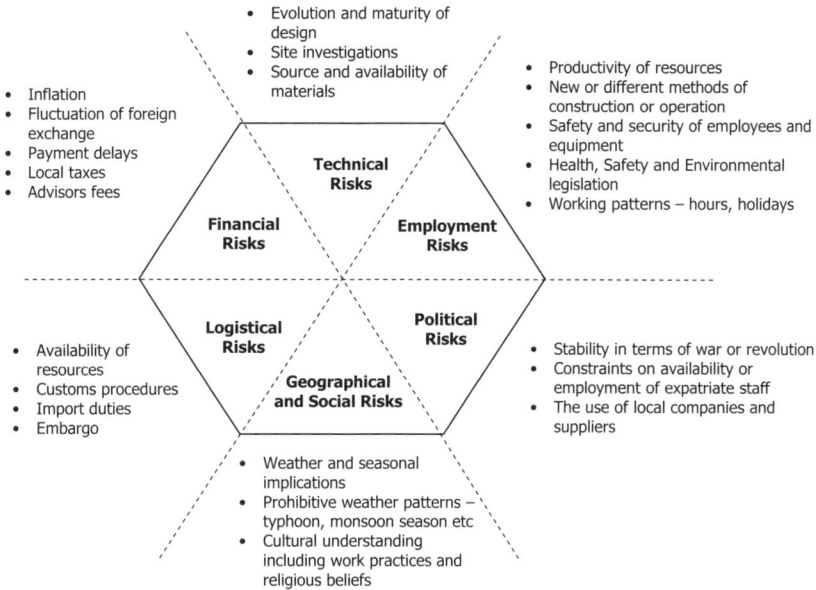

- Evolution and maturity of design
- Site investigations
- Source and availability of materials

- Inflation
- Fluctuation of foreign exchange
- Payment delays
- Local taxes
- Advisors fees

- Productivity of resources
- New or different methods of construction or operation
- Safety and security of employees and equipment
- Health, Safety and Environmental legislation
- Working patterns – hours, holidays

Technical Risks

Financial Risks

Employment Risks

Logistical Risks

Political Risks

- Availability of resources
- Customs procedures
- Import duties
- Embargo

Geographical and Social Risks

- Stability in terms of war or revolution
- Constraints on availability or employment of expatriate staff
- The use of local companies and suppliers

- Weather and seasonal implications
- Prohibitive weather patterns – typhoon, monsoon season etc
- Cultural understanding including work practices and religious beliefs

Figure 5.3   Areas of international risks

Source: Model developed from text *The Importance of Pre-Contract Risk Assessment and Management in PFI International Projects*, Professor S.A. Burtonshaw-Gunn (2008), from *Supply Chain Risk: A Handbook of Assessment, Management and Performance*.

financiers and insurance companies. An almost inevitable requirement on such international projects will be the provision of a guarantee to the lenders to cover their investment in the project prior its commencement. A significant guarantee to reduce the financial risk on public construction projects is often provided by a government to support operations in case of lower than expected revenues. However such guarantees are usually only available after a demonstration of the operating characteristics in line with specific design requirements. Typical areas of international risks are shown in Figure 5.3; these risks need to be considered not just during the project but importantly at the pre-contract award stage. On contract award other technical and project risks will also be identified and will have to be managed as described in the previous chapters.

It should be noted that the prime contractor's offering of a single point of responsibility naturally increases its own level of risk exposure; this itself may lie outside the organization's governance and the level of risk aversion of its organization. These risks may arise from construction risks particularly those around increased costs and project time extensions which itself may attract a cost penalty; and from operational risk where long-term project facility operation is required before transfer to the ultimate client. As first point of contact the prime contractor is *ipso facto* the first point of liability for claims by, and amongst, the preferred supply chain members and the client organization – be this private or public.

The final phase of the project will be covered by an operations contract covering the financial charges of the facility. PFI project risks are thus better identified and addressed when the interrelated financial, planning, construction, commissioning and operational risks assessments are brought together under a common risk management process under the control of the prime contractor.

On completion of a PFI contract, the facility is transferred to the public sector in accordance with the financial contract and operating agreement usually at no cost to the end user. In the event that the state wishes to take possession of the facility earlier than the contracted transfer date then some financial compensation will often need to be agreed and paid to the investor. Ideally all PFI projects will have been fully paid for by the time of transfer in which case the state may wish not to charge the project users anymore. This is more likely to be the case with PFI road projects and bridges than with a more technological facility such as a dock or an airport which will continue to require ongoing plant and equipment maintenance and have its own investment programme to allow it to conform to international safety requirements for its continued operation.

There are clear advantages for clients to offer projects using a Private Finance Initiate approach including the opportunity for third-party funding, risk sharing or even risk transfer. For client organizations (usually governments) they can also raise money from the letting of PFI contracts and at the same time preserve their own funds for other uses. Depending on the contractual arrangement such PFI-funded assets will ultimately return to state ownership after an agreed contract period. Prime contractors will have to determine if the project attractiveness is acceptable to them and then consider the risks of the venture. In addition such initial consideration may also take into account their wish to do business in the particular region or country; the projected costs and anticipated return on their investment; the marketing value of undertaking work in the particular country or with respect to other potential projects and the amount that it can utilize its own products and services or those of its known supply chain partners. Furthermore, it is suggested that the following risks will need to be considered and addressed:

- The **financial viability** of the project which will need to be undertaken on a sound business plan.

- The **limitations of the PFI investing consortium** in the event that it may not be able to influence design; this may already have been done with the risk of incorrect or inappropriate design features.

- The **risk** that the level of the traffic or passenger forecast, which represents the income potential and on which design is based, may be flawed or over optimistic.

- The **political stability and support** for third-party operation of what may publicly be viewed as a 'state' asset.

- The selection of **project location** – country, areas, regional development and so on – and if the prime contractor wishes to work in this location.

- The **influence of other government initiatives** or development schemes together with any changes in state legislation where this may impact on foreign investment and ownership.

- How such **changes** may impact on the project.

- The risk of changes in **environmental legislation** which may increase the risk to the operating assumptions made at the start of the project, for example road bridge usage or airport operation.

- Agreeing the right **contractual framework** and the need by the prime contractor to have position of 'influencer' – that is, regarded as a major shareholder not just a construction contractor.

- The risk that the prime contractor's **supply chain partners** may not be acceptable to the rest of the consortium or local government representatives.

- The risk that **poor performance** of the supply chain can impact on the projected SPV profits.

- The risk of pressure to use **local suppliers** proposed by host government.

- The risk that **other suppliers** than those preferred by the prime contractor may impact on financial arrangements and overall reputation.

- The risk that the **(SPV) consortium** will only be interested in the share of profits and not other long-term business opportunities.

- With high-profile projects, the risk that **good performance** will need to be demonstrated or else this will impact on other opportunities in the country or other PFI projects in other locations for different clients.

There will be ongoing project risks following the award of the contract, however for all parties considering undertaking a project in the international arena the status of the local economy and political stability are crucial for any project to be successful. Indeed, if contractors perceive that the risks are too high and companies are not interested in the project then it may only be possible for the development to be undertaken on a more traditional financial basis with government funding.

As an increasing number of PFI projects come to the market there is a view that many contractors offer a very conservative approach to design and construction in order to reduce both risks and costs and that innovative ideas are only being used when they can contribute to make the facility more profitable in the long term. Whilst some would argue that establishing a PFI project is a complicated process due to the number of different contracts, SPVs and the number of organizations involved, this need not necessarily result in a more expensive project when efficient design, construction and operation techniques are used and when the project is part of a series of similar facilities, that is, schools, prisons, roads and so on. Whether such prime contracting projects are offered as part of a series of projects or they are effectively 'one-off' facilities, as in the case of port and airport developments for example, the use of this form of procurement continues to grow in its popularity and use. As such the inherent risk with it emphasises the need for systematic and professional risk management especially as seen when the project is located in a country other than the contractor's home base.

PART 2

Financial Management

Financing of Construction Projects

CHAPTER

6

Basic Economic and Financial Principles of Construction Project Investments

This first chapter in the second half of this book commences with a look at investment in construction projects. In this case the term 'investment' should be regarded as a financial involvement in a project undertaking, aimed at gaining a benefit in the future; usually profit, but not necessarily so. This said, it is significant that the value of such investment expenditure or financial involvement which takes place is a known entity while the future potential benefits, that is, the profit from the investment, is only a possibility and not therefore secure.

By definition 'to invest' means to designate financial means not for consumption, but for specific profits to be earned in the future. On this basis, investment *per se* may occur in a number of ways; for instance through depositing money in a bank, buying bonds, stocks, machines or real estate property. As mentioned above the future benefits are not secured; therefore, every financial investment must be associated with an element of risk which must be analyzed as thoroughly as possible in order to eliminate its negative consequences. In order to plan the future investment or project undertaking effectively, it is often necessary to consider many various aspects of the investment activity. Such analysis is aimed not just at finding the current and future possibilities of the investment's performance, but also the threats to those projects.

Understanding and using the tools of financial assessment of investment undertakings is a natural element of the economic education of every investor regardless of where the funds may be invested. The investment process itself can encompass the entire project development activity and usually consists of the three phases:

- Phase 1 is the Pre-Investment or Pre-Decision Phase before the award of a contract, constituting the necessary tests, studies and expert's opinions aimed at selecting the best version of the project and making a decision on an investment opportunity.

- Phase 2 is the Investment Phase which consists of planning and contracting for the material deliveries, services, plant, equipment and furnishing. For construction projects this also includes the traditional activities of hiring the contractors and their supply chain for design, construction, installation and commissioning works.

- Phase 3 is the Operations Phase which covers the normal use of the facility as a result of its need for management resources, maintenance and repair requirements. This phase is of most benefit to projects where the investor has a long-term commitment other than just the delivery of the project, for example: in undertaking contracts such as PPP projects as described later in Chapter 8.

Incidentally these three phases were also mentioned in Chapter 5 with a particular focus on the project risk considerations at the pre-investment phase for international projects. The development of an investment project using the three phases is shown in Figures 6.1, 6.2 and 6.3 with a discussion on each. It is important to note that the investment decisions of the three phases are considered and used to secure the project's investment at the outset; whilst in the three phases described below, the funding obligation will need to be allocated in *toto* at the outset.

Phase 1: Pre-Investment

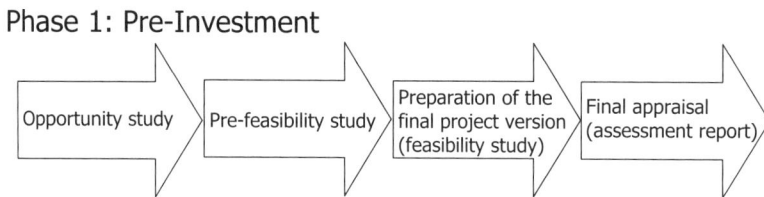

Opportunity study → Pre-feasibility study → Preparation of the final project version (feasibility study) → Final appraisal (assessment report)

Figure 6.1   Phase 1 process for project investment

The four stages of the pre-investment phase provide:

1. A study and analysis of the project concept together with initial advice on the investment and the preliminary economic assessment of the project (opportunity study).

2. A preliminary study covering the project selection and specification of the method of construction and project implementation (pre-feasibility study).

3.   Formulating of the final version of the project (feasibility study), including a complete set of technical, economic, trade and financial aspects of the investment.

4.   Final assessment of the project (assessment report) as a basis for the implementation decision.

As the pre-investment phase is divided into the above four stages this avoids, or at least reduces, direct transition from the concept to a final project study without examination and presentation of any alternative solutions. The first stage of an opportunity study prompts the potential investor to consider various investment options and proposals for projects that are to be subject to analysis during the further stages. The aim of analysis in the first place, is a quick and inexpensive general presentation of the most important aspects of the investment, particularly the project viability and the potential profitability of the investment made. This leads to the next step of considering a more detailed analysis. The pre-feasibility study of the investment project is aimed at analyzing further the project viability and its profitability, this stage is often not overly costly or time-consuming as it builds on previous work. It should however allow the selection of a concept and suggest a final version of the project. Therefore, for maximum benefits, the structure of the pre-feasibility study must be consistent with the final objectives of the project.

In some situations, additional comparative analyses may be conducted, pertaining, for instance, to the selected aspects of an investment project with this likely to cover a market analysis, aimed at specifying the demand for the product to be manufactured, material resources needed, the quality and quantity of production, plant location, selection of equipment and tools and so on. These are usually only prepared in the case of large projects especially if the investor is required to have an operational role or receive financial return on the investment from the performance of the constructed facility such as with transport, healthcare and education facility projects undertaken on a long-term contractual PPP arrangement.

The project feasibility study, or any other form of analysis such as a project business plan, should serve as a basis for the implementation decision with regard to the technical and technological issues, as well as economic, organizational, practical construction and financial considerations. After completion of the feasibility study, the parties interested in the project conduct an assessment according to their own objectives, early risk assessments and, cost and benefit criteria which is then documented in a final assessment report. Naturally the more thorough the pre-investment study, the easier it is to assess the project and to make the final investment decision on launching a realistic and achievable project.

Moving to the investment phase shown in Figure 6.2, which consists of a wide range of consulting, construction, commercial and technical management activities, the first stage would reveal any problems that need to be encountered for the purchase of land, through to the outline and then the detailed technical design and contracting strategy. This first stage will also include involvement in production of tender documents, assessment of offers, contractor negotiations prior to contract award for technical design, construction, installation of equipment and commissioning activities up to recruitment and training of operational staff. This phase is expected to be of a significant duration necessary to cover the detailed technical design, preparatory works on site and preparation of schedules of construction works and plans. The phase commences with signing of contracts between the investor and the financial institutions, consultants, architects, contractors, various suppliers of materials and so on.

Phase 2: Investment Phase

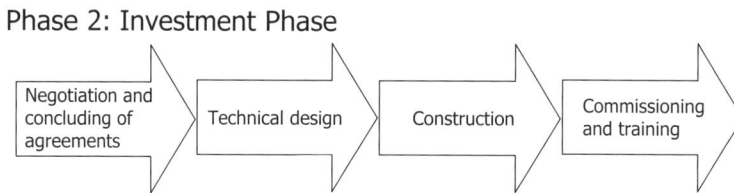

Figure 6.2   Phase 2 process for project investment

The five stages in the investment phase provide:

1.   Preparation of the technical report and the implementation plan.

2.   Negotiating and concluding contractual agreements for suppliers, works and services, construction or modernization of the facility (design, construction, installation, commissioning of plant and process equipment).

3.   Obtaining of permission for spatial development and performance and operation of facility.

4.   Delivery of facility (in whole or in part) for operation.

5.   Management of the performance of the project, including the supervision, control and coordination of performance of the facility during this stage.

Whilst the construction activities include the design, on-site and off-site fabrication, building and commissioning, it should be noted that from a financial viewpoint this must also include the management of finances in accordance with the financial plans and supervision of the plan performance. Particular emphasis will be placed on systematic supervision of the material development and compliance of the investment project with any schedule milestone payment dates, taking into account the extent of any deviations. Often a problem associated with investment in the operations phase is in the time and effort to monitor task completion with regard to the technical and financial aspects as well as accounting for risk funding expenditure.

Finally the phase 3 stages cover the operations phase of the investment process where the necessary compliance with the technical and economic parameters will need to be documented. These values will be decisive for the size, quality, timeliness and costs of constructing the facilities and, in some investment circumstances, in the provision of services and other project delivery arrangements.

Phase 3: Operations Phase

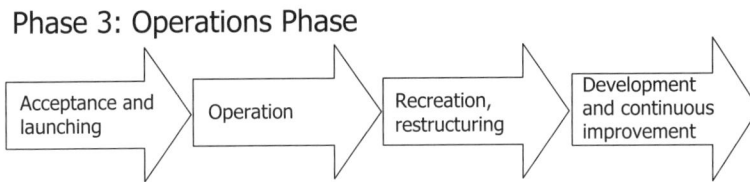

Figure 6.3   Phase 3 process for project investment

The above five stage operations phase identifies:

- The formal completion of the project with handover to the client (as appropriate) on completion of the commissioning and acceptance testing.

- The need for the provision of training for employees, improvement of use of tools, processes and equipment and so on.

- The official commencement of normal operational use of the facility and the start of continuous improvement over the life of the facility.

This third phase usually commences with the acceptance, that is, the physical completion of the project and the settlement of the contractual obligations. At the same time, statutory warranties with regard to any defects and the appropriate securing of effective operation of the facility performed by the main contractor will commence aimed at warranting the functioning of the facility with respect to its technical, economic, financial and practical aspects. In the initial period of launching of the production, operation or service activities, problems may arise which are normally associated with such issues as the application of production technologies, use of equipment and so on. Most of these will arise as a result of previous investment decisions made on the project which may have been a compromise between the client requirements and the amount of investment funding available. From the long-term perspective, if the study applied in the pre-investment phase turns out to be erroneous and potential defects are identified in the operations phase, corrective actions may turn out to be difficult to rectify, as well as very costly to change. The relationship between decision making on a construction project and the level of influence is discussed further in this chapter.

General Criteria in Making Investment Decisions

An investor's organization in making investment decisions often engages substantial financial resources and assumes the associated risks in order to make a well-founded decision, in general the investor – whether individual, company or consortium – must consider and determine the following:

- Firstly, that the objectives of the investor are defined precisely and agreed upon with the appropriate bodies approving valuations at the earliest stage, since all of the subsequent activities will be focused on achieving these objectives in the most efficient way. The main project management variables of cost, time and quality may conflict with each other and thus it will be extremely important that the team implementing the venture knows the relative priorities, for instance, the minimum completion time, the required quality standard and the maximum project cost. These objectives are rarely coherent, and as such will influence substantially the valuation, attractiveness of the project and its implementation activities.

- Secondly, the commercial environment in which the facility is to be constructed and then used will require market intelligence as it will be necessary to analyze and predict the market and economic trends, as well as technological progress and the activities of any identified competition. This will allow the competitive risks of the investment to be properly considered.

- Thirdly, is a requirement for realistic estimates and forecasts which will need to be made at the pre-investment stage. These are applicable throughout the entire cycle of implementation and use of the project. As a result, the cost budgets prepared on the basis of incomplete documentation will be misleading as they will sensibly include a risk contingency allowance for unknown risks.

- Next, is the need for extensive studies on the risk associated with investment specifically with an indication of the level of confidence of the estimate, as well as resources required to cater for any unexpected expenditure. What is more important is that the risk assessment at this early stage of the project can be used to indicate the areas in which additional information will be required to provide an increased level of risk confidence. Sometimes the risk assessment will also make it possible to plan a response to problems that may arise, thus reducing the risk level of the project.

- Finally, a project plan will need to be produced showing the most effective manner of implementation of the project and the achievement of its objectives, taking into account all of the limitations, constraints and the risks involved. Such a project plan defines the appropriate contract strategy (see also Chapter 8) and will include a programme showing the deadlines for making of key decisions and concluding the contract agreements.

There is a general belief that success of a construction project depends to a large extent on the expenditure in the pre-investment stage preceding the approval and any contract award. However, a conflict arises here between the need to gather as much information as possible to reduce the risk by the 4Ts (as covered in Chapter 4) and to shorten the investment period, as well as awareness that the expenditure incurred in the pre-investment stage will be wasted if the venture is not approved.

In the case of a large-scale construction project, expenditures for the pre-investment stage, (shown as 'valuation' on Professor Smith's model of Figure 6.4 below) rarely exceed the level of 10 per cent of the project's capital expenditure as costs beyond this level are significantly difficult to recover. Changes in the costs planned for further stages of the investment process bring relatively low savings when it comes to reduction of costs of implementation of the project as illustrated in Figure 6.4.

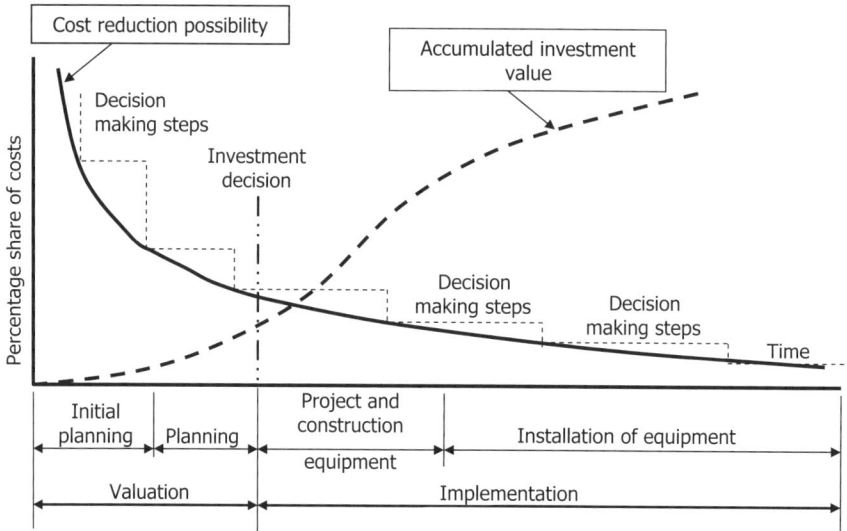

Figure 6.4   Percentage share of costs during the venture implementation cycle

Source: From '*Managing Risk in Construction Projects*', by Smith et al (2006). Reproduced with kind permission of Blackwell Publishing.

In practice the valuation of the project is a process of research, analyses and estimates conducted during defining the selected project and its alternatives. The objective of this study is to assist the investor in gathering information on which to make rational choices with regard to the nature and scale of investment within the framework of the project and providing assumptions for its construction. Project valuation is normally an iterative process which may be repeated when new concepts arise, when additional information becomes available or when uncertainty is decreased. This can prompt the investor to make the final decision on a project and encourage the investment on the basis of an expected return of the expenditure made.

The proposals based on these analyses should define the main parameters of the project, such as its location, the technology applied, the investment size, sources of financing, financial contracts, any specialist construction materials or features, and the market forecasts and estimates with regard to both expenditure and revenue from the investment. The investment decisions may be subject to a number of limitations due to non-financial factors, such as:

- the organization policy, strategy and objectives;

- the availability of resources, such as workforce, management staff or technology.

Defining a programme for implementation of a project usually means that works should be commenced as early as possible as there can be no income on which to derive profits until the investment is completed. Usually, market conditions or other liabilities enforce the final project schedule, none less so than for key showcase projects such as the Millennium Dome and 2012 Olympic sporting venues for example. However, it will be necessary to define the programme for the project valuation and implementation stages for two reasons; firstly the commencement date and the project costs may be defined, and secondly that the liabilities of the investor may be estimated and verified for the purpose of assessment of the investor's ability to cover these liabilities.

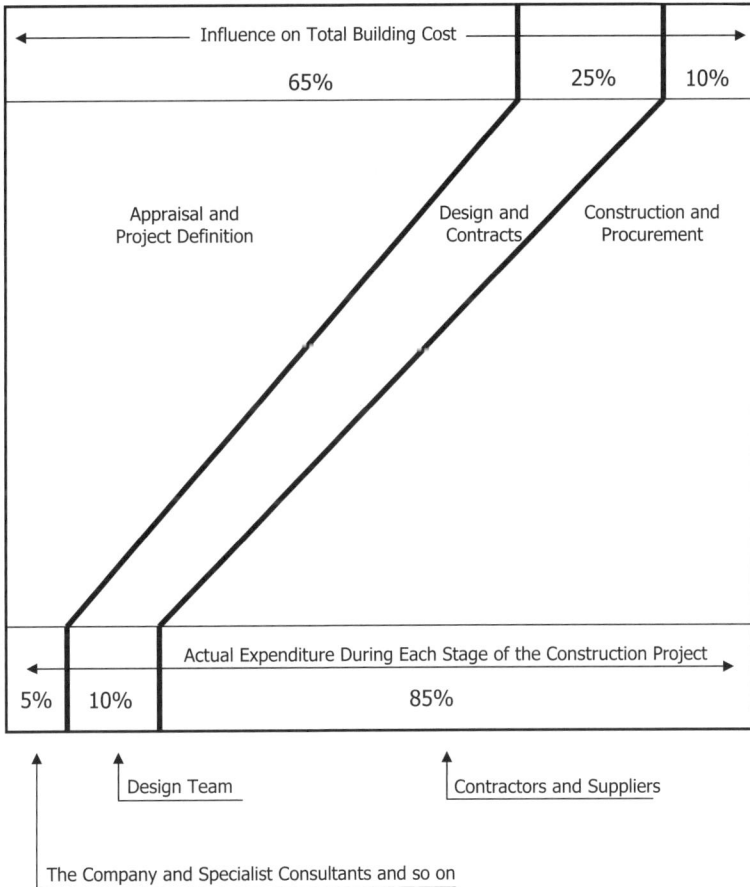

Figure 6.5   Influence of various parties on the cost of new facilities

The time factor should be taken into account as the valuation stage may be of significant duration and as many costs are time-dependent they may increase substantially in the case of any significant delay. Therefore, the project's construction programme must be realistic, and not over optimistic as a result of client or end-user pressure; this must be fully taken into consideration while determining the project objectives.

Finally, the influence of the parties on the cost of the construction activities needs to be understood and that, whilst the client organization may only be involved in the project at the early stage, the decisions made at this time play a substantial role in the overall project direction and its costs. Conversely whilst accounting for the majority of the project spend, the construction and procurement activities have comparatively little influence on the overall final project cost outcome as this is determined by the client requirements and compliance with the detail design. This relationship between the project parties and their relative influence of cost is shown in Figure 6.5. Accepting this relationship, it is also clear that the attention to risk management at the earliest stage provide the most economic advantage to the project costs where 'early fixes are cheap fixes'.

Management of Risk and Uncertainty of Project Investment

If an investment in a construction project is treated as a way to invest capital, apart from comparison of the expected profits, it will be necessary to take into account the possible risks which may inhibit the project's success and hence impact its financial performance. It is natural to compare the level of risk with other possible investment opportunities, such as the stock exchange, bonds or various currency accounts where such comparison often suggests to a more risk adverse investor to spread their investment capital in various segments of the market to provide a range of options and general versatility. These decisions protect investors, whether individuals, a group or an organization, against fluctuations in a single segment of the market (for instance, stagnation in the housing sector witnessed in 2008 or downturns in the commercial or public sectors), bankruptcy of a subcontractor or problems encountered during implementation of a single investment project.

Each of the individuals or groups making an investment assesses the risk associated with the given opportunity against the expected benefits. This should focus on the basic question: 'Can the investor afford a potential loss for the sake of a potential profit; or in simple terms is the profit worth the risk?'

The two basic concepts associated with risk decision making are uncertainty and risk where uncertainty is the lack of knowledge and inability to assess the probability of emergence of a given phenomenon. The concept of threat is associated with a specific risk of its emergence, if it is possible to assess, estimate or calculate the probability of its emergence, and the actions that this may require.

As shown in the earlier risk management chapters, both uncertainty and threat need to be managed by a systematic risk management process, recorded in the risk register and periodical reviewed and updated as the construction project is undertaken.

In general it is possible to say that investors fall into three categories where each behaves differently depending on actual or changing circumstances. These three groups are:

- At one extreme are those that treat the possibility of profits or losses as a challenge, these may be termed as risk seekers. For this group, searching for risk provides a high level of satisfaction as it is inevitably associated with delivery of a high level of profit.

- At the other extreme are those who overestimate the potential losses and thus are less interested in the potential benefits, these may be categorized as risk adverse. This group of investors usually avoid risk as much as possible by demanding that the investment project opportunity brings certain profits, but also accept that these are likely to be at a lower level of profit than for those projects favoured by the risk takers group.

- A neutral approach is an intermediate stance where analysis of threats and the selection of projects are characterized by the acceptance of an average level of risk and a desire to obtain a moderate level of profit. As risk taking follows a statistical 'normal distribution' profile, the majority of investors fall into this category.

Whilst the attitude of individual investors towards risk may of course change as the circumstances change, the collective attitude towards risk – avoiding, searching for or attempting to stay neutral – remains much more consistent for organizations.

Methods of Raising Capital for Construction Projects

Where the client company or its appointed construction company does not have enough immediate funds from either its retained profits or its normal business income activities to provide sufficient funding for a large-scale project, the options to raise and/or increase initial project funding are discussed below. Within the chapter, four areas of project funding are examined as shown in Figure 6.6, within each area are a number of funding sources.

The Stock Exchange is aimed mostly at medium-sized and large organizations who are trying to obtain finance to develop their activity from investors willing to deposit their money in companies listed on the stock exchange based on the knowledge of their past financial performance, reputation and other company details available to investors. These investors should not be considered loyal to the company as when they deem necessary they can easily and quickly sell their shares – even if this means at low prices. This may result in some loss of reputation with implications for other investors, suppliers and clients. However the Stock Exchange supports the development of companies by making it easier for them to obtain funds for subsequent investment projects; this feature may also encourage foreign investors and is conducive to the privatization process.

Becoming listed on the stock exchange is often regarded as a major achievement for a company due to the new access of large-scale funds which may lead to new or accelerated company development. However, being listed on the Stock Exchange is a serious undertaking and not an easy task as the management board must meet rigorous conditions at the time of the Stock Exchange debut and in the subsequent years of their activity.

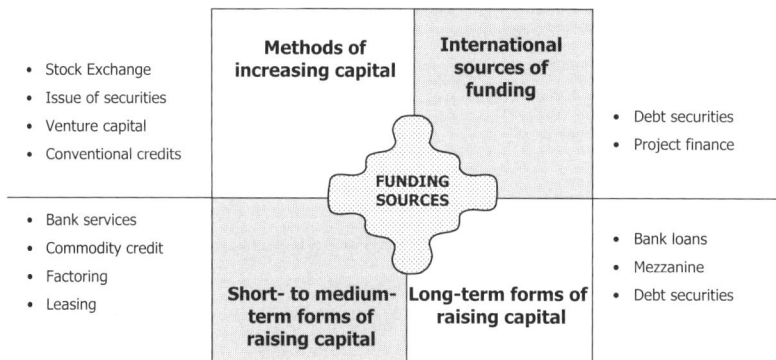

Figure 6.6   Sources of project funding

The initial capital of a company is clearly the investment that the company owners (stockholders) made during the issue of stocks through their purchase. Thus, the stocks are associated with a right to company shares, that is, part of its profits and the right to vote during the meetings of the management board. This enables the shareholders to make significant decisions concerning the company activity including views of its strategy or policy. Through each stock issue the company may increase its capital and thus obtain further financial means for its business activities.

On the other hand, debt securities, such as bonds, are another form of financing of activity and whilst these do not increase the initial capital they do allow for additional financial input. Debt securities do not provide the investor with any control in the company, although give the right to obtain interest as a return on the money invested.

From the perspective of the company, two basic advantages of regular stocks over bonds can be seen. Firstly, there is no obligation on the company to pay dividends. In the case of losses, it is not necessary to pay dividends to the shareholders as this would further inhibit the organization's financial standing. Secondly, the capital does not have to be repaid as there is no repayment date for stocks, and thus there are no situations of a sudden burdening of the cash flow which might result in financial damage or, at the extreme, the liquidation of the company.

Apart from ordinary stocks, preference stocks can also be issued which also increase the initial capital of the company, but with more limited rights such as the right to a dividend even if at a lower rate than ordinary stocks. It can be profitable to increase the company's capital by issuing preference stocks with no weakening of the management of the company as preference stocks have no voting rights and would not allow the investor to interfere with the control of the company.

Venture Capital financing pertains to private companies which will become profitable only after they develop, which is usually between 3 to 7 years. The cost of funding such uncertain investments is witnessed in the demanded high rate of return. Venture capital funding is usually aimed at innovative companies, often based on advanced technologies, associated with great hopes for profits being several times higher than the capital invested. They are naturally also associated with a high level of investment risk. Investments may also be made in ordinary companies, which have a good product and which have been successful in the market, but which require new capital for a development opportunity, increasing their production capacity, taking over new market niches or for the development of their sales networks.

Venture capitalists, through acquiring of shares in private companies, share with the companies the risk of business activity, and do not expect any additional warranties from them due to this fact. Usually, venture capital funds are not involved in direct company management; however, they almost always maintain control over the company by participation in the company supervisory boards. The best example of seeing companies request funding through this route is shown on the BBC television programme 'Dragon's Den' where venture capitalists consider investment opportunities in return for company shares together with a return on their investment and often, but not always, some influence on policy and business development decisions .

Where company funding is required, and the cost is too high in terms of company share ownership to make venture capital funding attractive, conventional financial credit can be raised through:

- A form of mortgage with payments calculated such that the combined capital and interest payment in each instalment is equal, while the capital and interest rates change. This method means that initially the share of capital repaid is low, but it increases progressively.

- A mortgage arrangement covering equal capital instalments where the amount of capital repaid is the same for each payment, and the total amount decreases with time as it consists of the fixed capital rate and interest on unpaid capital.

- A maturity date mortgage where the capital is repaid at the end of the loan period as a single amount. This form of funding is most appropriate for projects as it can be matched to the income profile especially where a termination bonus is paid on project completion.

Having discussed the above funding options there are risks associated even with these conventional repayment methods due to the uncertainty of cash inflows and outflows, as discussed later in budgeting and cost control in Chapter 9. Many projects require flexible repayment mechanisms to be successful due to the potential changes in cash flows. Indeed, project cash flows should be maintained between the worst and the best case scenario, so that the payments can be made even if a risk occurs.

The agreement between the creditor and the borrower is usually terminated if the creditor has information insufficient to estimate with confidence the result of the venture. In most cases, the creditor analyzes the project itself, and not just its assets, as a source of guarantee of payment. The key parameters taken into account by creditors at this stage are:

- The total project size as this substantially determines the amount of money needed and the efforts to raise the capital, the internal rate of return (covered in Chapter 7) and the project's entitlements.

- The critical dates when the capital investors expect to obtain revenues from their investments.

- Any key deadlines or important dates associated with financing of the project, for example stage payment details.

- The actual cost of the project credit, the amount required and the timescale in which the amounts used are the highest.

- If there are any warranties of debt repayment issued by the country in which the venture is implemented.

- Any time-dependent cash expenditures before the income is generated, again this will need to be compatible with the project cost and an agreed payment plan.

It is very important to realize that project financing may influence the conditions of implementation of the project more than those of the construction solutions or the actual construction costs. Indeed, many well-planned projects have failed, especially in the operation phase, when the income generated by the project activities is lower than that expected at the outset.

International Sources of Project Funding

The second quadrant of Figure 6.6 covers international sources of project funding. Predominately the privilege of large companies is access to international financial markets to obtain project funding which would not be available or preferential on the domestic market. The first of these options is from Eurocurrency Loans (the name euro has nothing to do with the common currency of a part of the EU region) which may be incurred outside the country of origin of their nominal currency (for instance, in American dollars outside the USA). These loans are not subject to control by the financial supervision institutions of specific states (such as The Federal Reserve of the United States of America). Eurocurrency loans offer two main advantages. Firstly, they are associated with fewer rules which reduce their costs and increase their flexibility; secondly, these loans allow the companies which obtain income in foreign currencies to repay the loans incurred in the same currencies saving exchange rate charges and eliminating the risk of higher costs from exchange rate fluctuations.

Foreign Bonds offer another funding route when they are introduced on a foreign market by the issuer. These bonds are nominated in the currency of the home market on which they are introduced and are subject to all regulations existing in this market. As the governing regulations may be quite rigid for the issuer and may limit its ability to raise capital, the attraction of foreign bonds is being replaced by eurobonds. These are a type of international bond which are sold and processed outside the jurisdiction of the state in the currency of which they are nominated. They are characterized by advantages similar to the eurocurrency loans and also allow the creditor to remain anonymous. Eurobonds should be differentiated from bonds in euros (nominated in the common currency and sold within the territory in which it is used). At present, for the sake of differentiation, the term 'international bonds' is often used, while the term euro has been reserved for the shared currency.

Project Finance is a type of financing of large undertakings, in which a specialized entity is established to implement and manage a project. Repayment of the debts incurred takes place from the income made by the project either during its operational lifetime or at set points in time. This type of financing is increasing in popularity for projects undertaken under the PPP model (see also Chapter 8). Theoretically, the investor takes into account the repayment and security aspects of the investment and how the profits are made from the assets owned by the entity implementing the project – very often in PPP arrangements this will be a government department.

Short- and Medium-Term Forms of Raising Capital through Debt

The differentiation between short- and medium-term financing is not a clear one. Usually, a period up to 1 year is considered to be short, while up to 5 or so years is considered as medium-term. Short- and medium-term forms of raising capital through debt are not less important than those previously described and sources of funding for such periods may be available through a number of routes described below.

For many companies, especially for the smaller ones, banking institutions remain the major source of financing. The first and often easiest option is through an overdraft facility which provides permission to withdraw from the account an amount exceeding the funds gathered. The period for which the money can be withdrawn in this manner varies and depends on the agreement signed with the bank. Usually this facility is for funds of less than 1 year. The interest is calculated for the amount by which the account limit is exceeded; often on a graduated scale. Funding of this kind is particularly useful for seasonal manufacturers. However, the main disadvantage of this kind of arrangement is that banks often reserve their right to cancel this arrangement virtually overnight which, of course, may result in a disaster for an indebted company. Having said this, banks rarely exercise this right as it exerts a negative influence of their business reputation among their current and potential clients.

The next form of short-term funding is commodity credit which can be used when the provider of specific goods or services agrees to postpone the deadline for payment. Due to such postponing, the relationship of purchase and sale turns into a credit liability. This solution is applicable when the purchaser does not have the sufficient financial means and the supplier cannot find purchasers ready to pay invoices quickly. Agreements for this kind of financing depend on the conditions negotiated by both parties.

Factoring is another form of financing in which a specialized entity (known as a 'factor') purchases invoices from the seller, for which a commission is charged. The money is paid to the service provider immediately, ensuring its financial liquidity, and the debtor makes a payment to the factor. In the case of this form of financing of transactions at least three entities are engaged: parties to the factoring agreement, the provider of goods or services, the factor (usually a banking institution or a specialized factoring company) and finally the debtor, that is, the purchaser of the goods or services.

The final option described is leasing which is a medium-term financing type, although in some cases leasing agreements may be undertaken over longer periods (in the case of the real estate for example), or, at the other extreme, short periods (such as leases on office equipment). The basic entities in a leasing agreement are the lessor (the financing party) and the lessee (the beneficiary). The lessor, being an owner of the goods, provides these to the lessee to use for a certain payment which is usually made in instalments. Each instalment includes the capital amount, which reflects the value of the product, and the interest amount, which is the remuneration of the lessor. Leasing may be competitive to a bank loan, especially due to taxation advantages that this type of funding offers and, as the owner of the goods remains so throughout the entire period of the agreement, it is common that interest rates are lower than those of a traditional bank loan. Three types of leasing agreement may be considered:

- An operating lease where the subject of lease is delivered to the lessee, usually for a period shorter than the period of its wear and tear. At all times it remains an asset of the lessor.

- A capital lease agreement which is similar to a loan agreement where the lessor includes the subject of lease in its property and makes depreciation write-offs which it is then able to treat as a 'cost' along with the interest element.

- A return lease arrangement which makes it is possible to sell the goods owned by the entity (such as a building) and then use it on the basis of a leasing agreement. The selling party is provided with funds and able to use the sold goods at all times. Such action, however, results in a liability on behalf of the lessor and it may reduce the overall flexibility of the company.

Long-Term Forms of Raising Capital

The final quadrant of Figure 6.6 covers long-term forms of raising capital which concludes this chapter. Long-term funding which typically covers investments beyond the medium term of 5 to 7 years, is where the financial markets offer practically unlimited possibilities of raising capital so that selection of the most appropriate form of financing of a project is no longer a simple decision. The basic possibilities of raising the required amount through debt, such as issue of debt securities or making credits, is discussed below.

Raising capital through debt has several advantages in comparison with equity financing (issue of shares). The basic one is the lower price of this kind of financing, which is not only due to lower cost of raising capital (cost of issue of bonds, credit processing fees and so on), but also due to the fact that investors are able to accept lower annual rates of return than they would be in the case of investing in shares of a given company. The reason is due to the lower risk taken by the investor in the case of investment through a credit. Also, for the indebted entity, in a situation of a market success it is not necessary to share with the creditor any large profits generated. Another advantage is the possibility of treating expenses associated with processing of the debt incurred as the costs of operation; this allows taxation advantages on these funds and therefore a reduction in value of the effective debt costs. However, the potential risks associated with indebtedness should be kept in mind. Creditors are able to secure themselves against insolvency; for instance, they may claim their right by taking over some of the goods owned by the company which may easily lead to serious losses or limit the flexibility of the debtor in making decisions for themselves.

An alternative to credits in the market of securities is the direct use of loans offered through bank loans. The bank grants a loan and the debtor repays it along with the interest rates throughout the repayment period. Incurring of bank loans may be attractive due to the:

- relatively low administrative costs, lack of costs associated with advertising and issue;

- speed that the agreement between the bank and the debtor can be concluded and the loan itself can be granted very quickly – sometimes within hours;

- flexibility – in the case of any problems with repayment, banks are usually more eager to negotiate the repayment conditions than investors purchasing debt securities; in addition it is often easier to negotiate with one entity than with several.

There is a cost to arranging bank loans as the debtor may be required to pay some administrative fees in association with being granted a loan, this may be for example an amount of 1 per cent of the loan amount; however, sometimes this fee level itself can be subject to negotiation. On agreement to grant the loan, the company may have to provide appropriate security in the form of assets owned so that if it becomes insolvent it would be possible for the lender to regain some of its funds. Finally, in considering such long-term funding the deadline and schedule of future repayments should be negotiated carefully as it will be important to avoid a situation where a company engaged in a long-term investment has to make repayments of a large credit throughout the project investment period.

Financial Assessment and Performance of Projects

7

Financial Assessment of Project Investment

The traditional method of analysis of the basic construction project centres on the three project performance parameters of time, cost and quality. The interrelationship between these is well known in both the construction industry and in the project management profession and is shown in Figure 7.1. However, from an investment analysis, instead of a project management stance, the three project performance parameters of time, cost and quality are discussed below.

The cost of the project undertaking depends on the unit costs of many factors. These costs may change, often increasing and seldom reducing, depending on both supply and demand relationship and on economic trends particularly for long-term project durations.

As seen in the previous chapters, the process of risk management may lead to a change in the cost level with regard to work, equipment and construction materials. Similarly, the time of a construction project, which may be broken down into the design, construction, installation and commissioning activities necessary to complete the project may affect the financial performance of the project due to the time-related risks associated with undertaking each

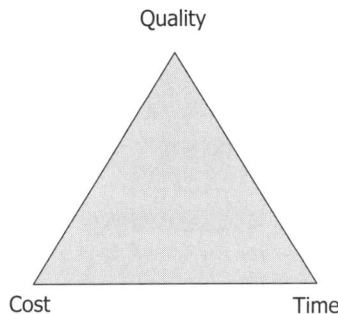

Figure 7.1   Triangle of project variables of time, cost and quality

of these activities. Within this assessment it will be necessary to include the procurement of any long lead time items and the general supply and demand for raw materials and so on. These programmes may not only affect the cost but also the availability, with consequences to part, or all, of the project programme. When looking at the requirement of quality, this is usually described in project's technical specifications and whilst these specifications can usually be met, the question is: in what time periods and at what cost? Hence the interrelationship of these three project and investment variables can again be witnessed.

All of the three mentioned parameters influence the profile of value of the project expected by the investors, however, it seems that the most significant of these parameters are time and cost. In addition it should be appreciated that it is extremely difficult to separate these two since, in the construction industry, the lengthening of the programme period will result in an increase of cost due to the addition cost of plant hire, temporary buildings and so on. However, reducing project time may also increase cost if additional resources are necessary to complete the project by an earlier programme date due to, for example, additional labour charges for unplanned overtime or shift working.

Moving to the parameter of quality, there may be a situation in which the construction facility is of high quality but fails to provide the necessary intended benefits for the owner. Therefore, instead of the term 'quality' the term 'quality profile' may be used to specify the organizational, technical and quality solutions, ensuring the benefit levels expected by the end-user, customer and investors. As such, it is proposed that devising a quality profile is a good method to describe the project's quality level required for the project and assigning an appropriate importance criteria weighting allows the relationship between the quality profile and the business objectives to be understood.

Meeting the three analyzed performance parameters shown in Figure 7.1 may provide a level of satisfaction of the investor's expectation if there are no deviations from the assumed expected factors characterizing these parameters. During the course of the project the value of each of the three parameters may change due to the emergence of risks which can significantly influence the importance of each. Indeed, changes in risk and uncertainty may affect the values of the time, cost and quality and as such Figure 7.1 is modified to reflect the additional and central role of risk now shown in Figure 7.2. As specific risks may be interrelated it is possible for these to result in an emergence of other risks or additionally change the proportions of their influence on the project. On this basis it may become necessary to revise the risk management strategy and the selected '4Ts' as previously discussed in Chapter 4.

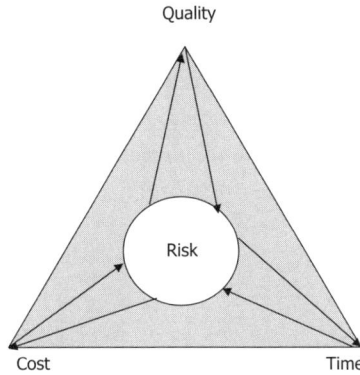

Figure 7.2   Triangle of project variables of time, cost, quality and risk

The correlation between risk and project value can be seen through analysis at the moment of emergence of a risk, when new risks may lower the final value of the project or, on the other hand, a reversed situation may occur, since it is possible for a new risk, or cancelling an already identified risk, to provide a project opportunity. Every construction investment is exposed to risks due to the diversity of the project activities which are rarely identical or even performed in identical conditions or similar locations.

Financial Performance of Projects

Financial decisions concerning projects are made by the owner, or a substitute investor on the owner's behalf, and whether an individual or group of investors, there will be a need for all projects to demonstrate their financial and economic rationality. Methods of project profitability assessment may be divided into ordinary (analyzing specified economic ratios) and discounted; although only the discounted methods are theoretically correct. These are generally applied by large corporations and international financial institutions such as the World Bank to assess the financial performance of large-scale construction projects.

The discounted methods allow for assessment of investment projects in a coherent and relatively integral way which can provide comparative information enabling the investor to select a project which most contributes to both the financial return on their investment and increases the company's market value.

In assessing the financial value of a construction project an appraisal may be undertaken using a number of techniques. A number of these are discussed below.

The first of the techniques is called Discounted Cash Flow (DCF) which in essence assesses the time value of money, such that it recognizes for example that £100 is worth more today than £100 in the future and therefore stresses that *when* the investors receive their money is just as important as *how much*. DCF can be used to forward project future values as shown in the example value projections in Figure 7.3 using the formula:

Value = Amount × $(1+r)^n$ where r = Interest rate and n = Number of periods

Interest Rate %	Year					
	0	1	2	3	4	5
5	100.00	105.00	110.25	115.76	121.55	127.63
10	100.00	110.00	121.00	133.10	146.41	161.05
15	100.00	115.00	132.25	152.09	174.90	201.14
20	100.00	120.00	144.00	172.80	207.36	248.83

Figure 7.3   Forward projections at different interest rates

In addition the DCF technique can also be used to discount back project values as shown in Figure 7.4. This method uses the formula:

Value = Amount / $(1+r)^n$ where r = Interest rate and n = Number of periods

Interest Rate %	Year					
	0	1	2	3	4	5
5	100.00	95.24	90.70	86.38	82.27	78.35
10	100.00	90.91	82.64	75.13	68.30	62.09
15	100.00	86.96	75.61	65.75	57.18	49.72
20	100.00	83.33	69.44	57.87	48.22	40.10

Figure 7.4   Discounting back at different interest rates

Another technique is known as Payback and Payback Period Assessment which examines how much money the project will make (or lose) where payback equates to the income less cost. In this way the payback period refers to how long it will take to recover the initial investment from the cumulative cash flow. Whilst this approach is quick and simple to understand, it does not account for the 'time value' of money as covered in the DCF technique above and can often be misleading and meaningless for the purpose of investment decisions. Additionally it does not account for the timing of cash flows and ignores all cash flows after the payback date.

Combining the DCF with the payback period provides a method called Discounted Payback Time which takes into account the influence of time on the value of the cash as a result of the completed project, but only in the assumed, expected payback time. Although this method still does not include revenue income after the accepted payback time it does include the elements of alternative cost and risk.

Profitability of invested capital is a relatively popular ratio for the assessment of an investment project. This is calculated by division of the sum of the forecasted net income, after deducting depreciation and taxation costs, by the investment value. This ratio is based on accounting income and omits distribution of cash flows during the investment itself. As such the result depends on the accounting calculation and not on the actual cash flows. The ratio ignores alternative capital costs and actual cash flows, as well as the risk of an investment but provides a guide by using current profitability as a reference point for investment selection.

Another assessment method covers the Return on Investment (ROI) which may be expressed as a percentage or fraction or ratio, for example, ROI 10 per cent, 1/10 and 1:10. The method provides information in relation of the income to the size of the outlay. Similar to the payback, this approach is quick and simple to understand although also fails to account for the time value of money. As with the payback method not only can it be meaningless for the purpose of making informed investment decisions, the average annual return can be particularly misleading as it does not account for compounding year on year.

The next assessment method is Net Present Value (NPV) which is the sum of the present values of all anticipated cash flows including the initial outlay. This then uses the DCF method to obtain the present value of all future cash income. This approach determines what the project is worth and whether a project opportunity is able to earn more or less than the investors desired rate of return. NPV expresses the profitability of an investment at the moment of the assessment. NPV is calculated by deducting the value of incurred investment expenses from the discounted value of the future income. Therefore, investors are able to consider a range of possible investments and then make a comparative selection of the most beneficial one.

Using the NPV formula, a development project is practicable if the calculation shows that the NPV is greater than zero, which indicates the profitability ratio of the investment. By accepting a project with a positive

NPV value, the company obtains cash income when present value exceeds the updated value of investment expenses. A negative NPV value indicates that the income value will be lower than expected, therefore, the project will be unlikely to meet the criteria of the assumed profitability at the given level of risk and its understanding will not be beneficial to the investors.

Whilst there are often a range of specialist or modified assessment methods, the final assessment covered here is called Internal Rate of Return (IRR). This uses a discount rate to enable the calculation for NPV equal to zero and is one of the most popular indexes for assessment in investment terms. IRR is calculated by an iterative trail and error process. It is less easy to use than NPV but when used well allows comparisons of competing projects to be made. Internal rate of return is the maximum percentage rate of a credit that would be practicable for the investor if all the expenditures were financed by a credit or loan. This method also calculates the actual return ratio from an investment.

Value Management Analysis

Value Management (VM) is the name given to a process in which the functional benefits of a project are made explicit and appraised consistent with a value system determined by the client. The need for VM arose from greater globalization in other industries and increased competitive pressures between companies, prompting recognition of a need for greater efficiency and effectiveness of companies. This led to developing VM as a tool which would control and improve effectiveness as the traditional approach – cost management – focused on efficiency, rather than effectiveness. As such, VM became a process undertaken with both client, the project builder and key suppliers. In the case of construction projects the common vehicle to this interaction between the client and the contractor is the contract. Both sides agree on the subject of what is to be built and on the terms and conditions under which it should be done. The VM approach is then about systematically aiding the process of design evolution of the project to achieve the best value for the client whilst preserving specified cost limits.

There are three characteristics which separate VM from a traditional accounting approach, these are:

- The requirement for a positive and pro-active approach through the use of a multi-disciplinary team-oriented creative process to generate alternatives to the existing solution.

- The use of a structured systems method.

- The relationship between function with value.

In the Value Analysis method the term 'value' is defined from the perspective of the expected benefits of the investor from the construction investment undertaking. These benefits are analyzed from the perspective of the entire project life cycle. Value analysis is therefore used to define the investor's expectations associated with the project, so that the investor enjoys the greatest possible benefits. As such VM is a method of maximizing the final effect of the investment according to the scale defined by the investor. Analysis of the value of the project should not be confused with the use of cost reduction or minimization of the project. VM goes beyond the ordinary reduction in costs although as a consequence of the VM analysis this method may also lead to a reduction of cost of the investment undertaking.

It is proposed that VM provides the following benefits:

- A method for defining projects in a clear way and in the context of long-term business needs of the investor and the end user.

- It provides support to the key processes of decision making on the basis of maximization of value.

- It encourages the generation of innovative solutions.

- It optimizes the balance between the initial design and the construction, maintenance and operation costs in the project.

- It provides a tool for measurement of value and allows for checking whether the ratio of value to price has been maintained.

In recent years VM has also evolved towards ensuring quality control which may be regarded as conformity with the specifications with an emphasis on meeting the expectations of the investor. The general value assessment typically includes performance, time, technology and reputation. At the same time, the scope of the method was broadened to include not only the products, but also the processes. In the construction industry a key issue in the field of VM has become the definition and measurement of the type 'value' in a manner which can enable the design teams to produce the most efficient solution for the required investment with regard to architecture, aesthetics, construction design, technology, maintainability and whole life costs. In the context of construction projects a measure of value can be expressed by the ratio of meeting the investor's expectations and the value of resources designated. In short it can be assumed that this relationship may be expressed as a ratio of values and benefits to financial expenditure for the project. Where as defined in the British Standard 12973:2000:

$$value = \frac{benefits\ gained}{resources\ used}$$

Having said this there are also clearly numerous instances where the investor's short-term view is in direct opposition with the end-users requirements. This often arises when the investor is looking for the earliest return on their investment, possibly at the expense of long-term maintenance cost savings to be funded by the end-user in the future – and long after the investor has left the project.

Tools and Techniques used in Value Management

The VM method is based on the specification of the function of the project planned. A typical feature of VM is the fact that the attempt to find better solutions by the appointed team is encompassed by a logical and sequential value analysis plan. The application of VM consists of devising answers to the three fundamental following questions:

1.   What is to be done?

2.   What are the investor's needs and expectations?

3.   What are the desirable features of the project?

In examining these questions, all available information on the project is gathered to define the functions of the individual parts of the project by creating clear and understandable statements. The information that needs defining during this process usually includes:

- The client's needs which cover the basic business requirements which must be met by the project so that its most important business or social objectives are achieved. The client's needs should not be perceived only in terms of usability since there may be other requirements, such as the need for good appearance or creating or promoting a positive image of the company.

- The client's wishes will need to be accommodated although not at the expense of the investor's return of investment.

- Information on any limitations of the project such as factors which impose certain requirements upon the project, environmental conditions, legal provisions and so on.

- The financial limits on the project expressed as the available amount of capital gathered at the beginning and throughout implementation of the project.

- The time spent on designing and building the facility and how this meets the client's requirements.

The Earned Value method is based on the measurement of performance results (also known as Performance-Based Management) and assessment of the results in relation to the financial resources used. With this technique, instead of subjective estimates based on individual viewpoints, earned value is based on a predetermined and quantifiable method for assigning value to completed portions of the project. As long as the predetermined method is used, the project's progress can be measured objectively and future projections can be made about the ultimate final outcome in terms of both cost and schedule performance.

Accurate project reporting is crucial for earned value as it takes the information collected on the tasks completed (hours spent, funds expended) and provides monetary assessments on the project's worth and its expected cost. The technique of earned value first came into popular use in the 1970s and is widely considered to be an excellent tool for quantifying project performance and relating this not just to the quality of the work done at a specified date, but also to the other two project elements of cost and programme, that is, time. One unique feature of Earned Value Analysis is the requirement for detailed 'bottom-up' planning. This feature is essential to the concept because it permits managers to measure performance at the level where the work is actually performed. As performance measurement information is collected and analyzed it allows a closer examination of the lower levels in the WBS to isolate and focus on their cost and schedule performance. It has this ability to direct management attention to specific or detailed areas with significant problems that makes Earned Value Management a useful and effective project management tool which can provide input into decision making and not simply as a cost reporting process.

The term Budget Cost of Work Performed (BCWP) is at the heart of the earned value performance measurement system. The BCWP is calculated by adding up the value of each work package that has been completed at a given

Budgeted Vs Actual

point in time. The measurement also needs to account for earned value of work packages that have been started but are still in progress. The difference between this measure and the Budgeted Costs of Work Schedule (BCWS) is that this includes the value of all work packages that should have been both completed and are in process at this point. Establishing BCWS and BCWP allows the project performance to be appreciated.

VALUE

The term ACWP is the amount spent to accomplish what has been done, that is, the Actual Cost for Work Performed, and is a traditional measure that covers all the costs incurred on the project up to a chosen point in time. This can be compared with the intended project plan spend commonly referred to as the BCWS.

The traditional method of controlling the project costs, in which the ACWP are compared with the planned costs (BCWP), does not deal directly with the scope of work performed. The control of costs and progress of work on the project is usually performed by various functional units, for their own purposes and according to their own controlling and reporting methods. Thus, the assessment indexes used by various departments, due to varying data structures, may not always be directly correlated. The traditional method of variance measurement (BCWP minus ACWP) fails to provide an objective view on the current state of the project (real costs being lower than planned at a

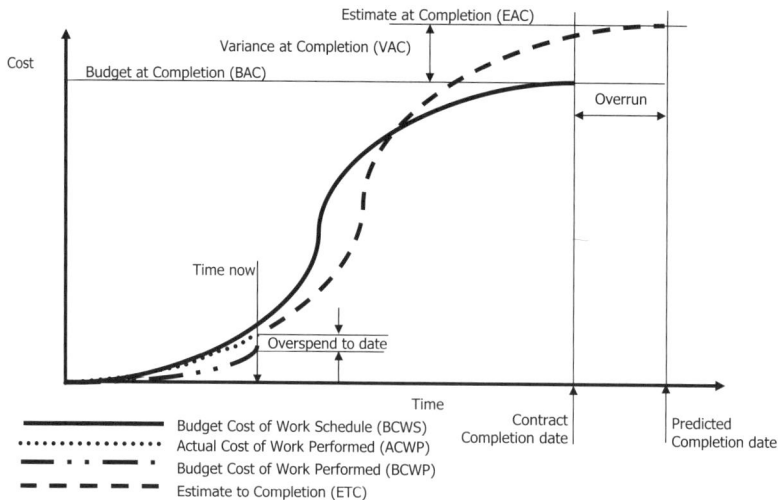

Figure 7.5   Project 'S' curves showing cost and time relationship

Source: From '*The Essential Management Toolbox*', S.A. Burtonshaw-Gunn (2008). © John Wiley and Sons. Reproduced with permission.

given point in time may simply be due to a failure to perform the planned scope of works) and cannot therefore serve as a basis for forecasting the future performance results.

The fundamental difference introduced by the earned value method takes into account the assessment of the project progress from three dimensions: the scope of works, the schedule and the costs incurred. This represents the planned value of the actually performed scope of works (BCWP) at a given point in time where application of the obtained BCWP value allows the cost variance measured in a traditional way to be divided into two independent variances: cost variance (CV) and schedule variance (SV) so that:

$$BCWS - ACWP = (BCWP - ACWP) - (BCWP - BCWS) = CV - SV$$

The last two indexes (the variances) are described as follows:

- The CV shows the deviation from the amount which should be paid for the work actually performed. The negative value of this index signifies that the work has been undertaken at a greater cost than that planned (for instance, because the initial assessment of the unit cost was too optimistic, inefficient or due to a poor initial estimate), while the positive value indicates that work performed costs less than originally planned. The fact of delaying or accelerating completion of the task (how much was supposed to be completed and how much has actually been completed) is not taken into account.

- The SV characterizes the deviation from the amount of work which was supposed to be completed at a given date, valued at the unit prices as planned (that is, budgeted). A negative value of this variance signifies that the task has been delayed whilst a positive value indicates that it is being carried out ahead of that planned. The real prices (costs) are not taken into account while calculating this variance. In order to perform an assessment of the risk of failure to meet the deadline it is necessary to take into account the project critical path and the time resources; as such SV is an additional measure of progress of the project completion schedule.

- BCWS - ACWP is a measured variance of the costs planned until a given point in time. It is of some value (when positive there will not be any problems with payments, if it is negative, such problems may arise) but it is not sufficient for a full analysis of the cost conditions for a given project activity. This calculation combines two aspects: firstly, a difference between the planned and actual costs and secondly, the difference between the planned and actual work progressed. This variance is not really useful for cost analysis and management but can be helpful in the allocation of funds.

Having described these variances it is not sufficient to limit the assessment of the actual condition of the project using the BCWP, CV and SV indexes. The Estimate cost At Completion (EAC) should be assessed to verify whether it is within the range of the budget at completion. The Budget At Completion (BAC) is the budget available to the project manager which does not include any reserve or contingency funds. Whilst the budget may be changed it should only be done so as a result of a decision of the appropriate investors or client representatives. When decisions are made, the BAC is modified from the opening budget where the following formula is always applicable:

BAC = Available (opening) budget + all accepted modifications

The EAC is the term for the cost estimate made at the beginning of the project. This value may change with time, when more knowledge of the project becomes available and a better assessment of the total cost of the task or project definition is known.

The term Estimate to Complete (ETC) means the planned expenses for a given task or project as of the current point in time.

The following calculations can also be made:

Estimate at Completion (EAC) = ACWP + ETC

Variation at Completion (VAC) = BAC - EAC

For assessment of the project progress and forecast of the final results the following two terms and calculations are also useful:

a)   The Cost Performance Index (CPI) = BCWP/ACWP

The Cost Performance Index (CPI) above indicates the relationship of the costs compared to the budget. A CPI of 0.80 will mean that for each £1.00 spent only 80 per cent was spent in accordance with the budget; in other words, for each £1.00 spent, work worth £0.80 was performed.

b)   Schedule Performance Index (SPI) = BCWP/BCWS

The Schedule Performance Index (SPI) calculation shows the relationship between the work planned and that which has been performed, so that an SPI of 0.9 would signify that for each £1.00 of work planned, only work worth £0.90 has been performed.

When the CPI and SPI indexes are equal to 1.0 everything is progressing as planned. When the SPI is below 1.0 this will indicate that the project has been delayed. Regardless of how the work is progressing, the SPI will eventually achieve the value of 1.0 because after performing all work the BCWP will be equal to the BCWS. When the SPI is applied together with the critical path method of project planning this allows for an objective assessment of the progress of the project work to be made with regard to its scope and the deadlines for completion. It should be stressed that the natural tendency of counteracting delays by assigning additional resources may improve the SPI but may also result in irreversible losses on the cost performance of the activity as funds spent on making up for the lost time are only very rarely recoverable.

Unlike the SPI, the CPI is not only subject to self-adjustment, but negative trends in the cost of the activity are hard or even impossible to reverse. Therefore, the project manager will have to focus on the tasks from the critical path, aiming at achieving a balance between the schedule performance (SPI, critical path and time reserves) and the cost performance (CPI).

Indexes of the earned value method may be determined for a given point in time or indeed for a given period. Practice has shown that for assessment of performance results it is better to apply accumulated values which are less sensitive to the anomalies and serve as a basis for assessment of trends in the project performance. For monitoring the remaining scope of work, the To Complete Performance Index (TCPI) may be used:

To Complete Performance Index (TCPI) = Work left/Funds left

Where:
the work left (the planned cost of work remaining) = BAC - BCWP,
and
funds left = BAC, or EAC (or New fund) minus ACWP.

The TCPI shows the cost performance that is required for the remaining scope of the project in order to maintain the project funding level available. On this basis a TCPI of 1.10 means that each £1.00 spent on the remaining works should bring work worth £1.10.

Integration of the project scope with the schedule and cost can be seen to provide a good level of measurement and assessment of: the actual performance results, the trends in performance and the forecasts of the future results. It may also be used as a basis on which project management decisions can be made. Use of the earned value method for project analysis will not prevent the project from exceeding its costs or failure to meet deadlines, however, it does provide an early indication which can trigger an alarm and a requirement for action. At the centre of the earned value analysis is the concept of earned value itself; simply stated this is a view of *'what we got for what we spent'*. For clarity, the terms and their associated formula are show in Figure 7.6 below.

Term	Meaning	Formula
% Complete	Percent complete (of the project's work)	BCWP/BAC
% Spent	Percent spent (of total budget)	ACWP/BAC
BAC	Budget at Completion	Cumulative BCWS
BCWS	Budgeted Costs of Work Schedule	
CEAC	Calculated Estimate at Completion	BAC/CPI
CPI	Cost Performance Index	BCWP/ACWP
CV	Cost Variance	BCWP - ACWP
Desired Margin	Desired margin	Revenue = Cost/1-margin
EAC	Estimate At Completion	ACWP + ETC
Equity	Equity	Assets - Liabilities
ETC	Estimate To Completion	CEAC-BCWP
GP	Gross Profit	Revenue - Costs
Income	Net income	Revenue - Costs - Expenses
Margin	Profit Margin	Gross Profit/Revenue
Markup	Markup	Revenue/Cost
Revenue	Revenue	Cost + Gross Profit
Revenue using cost	Revenue using cost	Cost + Gross Profit
Revenue using cost and expenses	Revenue using cost and expenses	(Cost + expenses) × markup, or (Cost + expenses)/1- desired margin
SPI	Schedule Performance Index	BCWP/BCWS
SV	Schedule Variance	BCWP - BCWS
TCPI	To Complete Performance Index	(BAC - BCWS)/(BAC or EAC - ACWP)
VAC	Variance at Completion	BAC - EAC

Figure 7.6   Formulae used to calculate earned value

Advances in Contract Strategy

Introduction to Contract Strategy

The term Contract Strategy is used to describe the policies for the organization relating to the provision of goods or services from other parties; this could for example cover equipment, design or construction and the policies for the associated contractual arrangements chosen for the execution of a specific project.

The development of a contract strategy is an important task for the client or the appointed project manager although often insufficient time and effort is given to this activity. It comprises a thorough assessment of the choices available for the implementation and management of goods, design and construction and usually follows a pattern of interrelated decisions which seek to maximize the likelihood of achievement of key project objectives. The selected strategy is likely to be optimal in that it will need to satisfy a variety of constraints and be sufficiently robust to withstand the uncertainty associated with a project such as a construction project undertaking.

Another reason for its importance is that the decisions taken during the development of a contract strategy affect the responsibilities of a number of parties; they influence the control of design, construction and commissioning and hence the coordination of the parties; they allocate risk and define policies for risk management. Finally, such decisions define the extent of control transferred to contractors and supply chain partners. On this basis, therefore, such contract strategy decisions can be seen to affect the main project variables of cost, time and quality and, on the 'softer' side, the relationship of the parties embraced by the contractual agreement.

The first step in the development of a contract strategy is to identify the main project parameters; in strategic management terms this is often referred to as 'strategic choice' and these are likely to cover:

- the project characteristics;

- the organizational system for the design and construction activities of the project;

- the type of contract;

- the tendering process.

Each of these parameters may be regarded as distinctive and internally coherent subsystems, within which there are numerous discrete choices. However, they are also interrelated as the optimal contract strategy will be one which displays a consistent integration of the selections across each of the above listed strategic areas. The main decision areas for each of the subsystems are discussed below:

Project characteristics	• The identification and ranking of primary objectives concerned with cost, time and functional performance. • An assessment of conflict between primary objectives and establishment of tolerance limits for at least one of them. • The identification and ranking of secondary objectives. • The identification of project constraints including priorities of timing and phasing. • An identification of project risk and overall risk management policy.
Organization of equipment supply, design, construction, installation and commissioning	• Selection of appropriate size and scope of contract work packages. • Choice of appropriate design organization and allocation of design and supervision responsibilities to the client, consultant(s) and contractor(s). • Selection of suitable contractual arrangement for contractors involved in design or material supply.
Type of contract	• Selection of appropriate type of contract from lump-sum, measurement, target cost or cost-reimbursable arrangement. • Selection of an appropriate mechanism for the measurement of work and evaluation of payment – especially for varied work.
Tendering process	• Method of appointment of design consultants, main contractors, suppliers and so on. • Method of appointment of subcontractors and second-tier suppliers. • Decision on whether prequalification is required. • The method of tender analysis. • Choice of suitable conditions of contract, assessment of appropriate provisions and modifications to any standard forms of contract if used.

When analyzing the choices, the client organization or its project manager must be cognisant of all the factors, including that of risk surrounding the project, which may influence this choice. Having decided on the organizations required for the project there are a number of choices available to the client and to the project manager for the management and execution of design and construction of facilities and for the provision of plant and equipment. Most projects require the services of organizations which are external to the client to undertake these functions. In practice, some organizational structures are closely linked with a particular type of contract, for example, the conventional approach with a measurement contract or fee contracting on a cost-reimbursable basis. However, this is not always the case and it is preferable to consider the decision of organizational structure as separate from, but interrelated with, the decision on the type of contract.

Having mentioned organizational structure it should be noted that these differ in the following key areas:

- The role of the parties.

- The management procedures and the emphasis given to the management of design and construction as these may be undertaken together by one organization or alternatively by two separate companies – one covering design and the other the construction, installation and commissioning activities.

- The allocation of risk

- The method of payment.

The logical method of choosing an organizational structure is to determine the approach which most closely responds to the project characteristics, (objectives, constraints, risks). This choice is facilitated by subdividing the overall decision into several subsidiary and interrelated decisions covering the following:

- the size and nature of the work packages within the project;

- the selection of the design team from in-house resources or external consultants/contractors (or some form of combination of these two);

- the method of management and coordination of the design teams;

- the extent to which construction is to be separated from, or integrated with, the design element;

- the procedure for managing the design/construction interface;

- the process of supervision of the construction activities;

- the benefits of, or restrictions on, using a combination of organizational structures within the project;

- the extent to which the Political, Economic, Social, Technology, Legal and Environmental (PESTLE) factors are dominant, for example in relation to the use of direct labour, labour-intensive construction and organizations of certain nationalities.

An initial assessment of the subsidiary decisions outlined above should lead the project manager to define an overall policy which will then need to be coordinated with the organizational structures offered by the industry.

In the construction industry the types of contract are primarily distinguished by their payment systems and four main types can be identified. These types are: Lump sum, Measurement, Cost-reimbursable and Target Cost, although generically it is possible aggregate these into the following two distinct classes as follows:

1. **Price based** – comprising lump sum and measurement Contracts where payment is based on prices or rates submitted by the contractor in their tender. These prices are deemed to include all costs, overheads, risk contingency and a profit element.

2. **Cost-based** – covering cost-reimbursable and target cost, contracts cover the actual costs incurred by the contractor which are reimbursed and in addition a fee is paid. The fee is deemed to include those costs which are not defined as reimbursable, plus overheads and profit.

It should be noted that within the industry the term 'type of contract' is quite commonly interchanged with the term 'form of contract'. The choice of type of contract is influenced by following:

- the size and nature of the individual construction or supply packages;

- the timescales associated with the issue of the tender documents and the required start, or more likely completion, of the construction project;

- the extent of design change which can be anticipated during construction;

- the appropriate allocation of risk between the parties, including suppliers, contractors and the client;

- the avoidance of incompatibility of the type of contract with others on the same project or at adjacent sites, that is, the choice may have already been previously made and can be repeated for this project;

- the compatibility of the choice with the resources which the client and its advisers are able to commit to the project;

- the appropriateness of the provision for flexibility, incentive and risk allocation.

This last factor is worthy of amplification since it is often central to the contract decision. In looking at incentive, the aim is simply to provide an adequate incentive for efficient performance by the contractor which is matched by an incentive for the client to provide the necessary information and support to the contractor in a timely manner. An ideal contract should enable the client to introduce a 'reasonable' number of changes which may be anticipated but not defined at the tender stage. An important requirement is that the contract should provide for their systematic and equitable evaluation. However, the initial degree of expectation of change is frequently exceeded in practice unless design changes are strictly controlled. The final point of risk allocation should be undertaken to allocate all risks between the client and the contractor, this should be based on a proper identification and assessment of their implications (see previous chapters) and must take into account of the management and control of the effects of risks which materialize.

It can be seen, therefore, that the selection of the contract strategy itself contributes to the ownership of risk: this risk associated with a Cost Plus contract rests with the customer where the supplier is not encouraged to save costs. In this way, however, the customer obtains the highest quality work. This approach is most applicable when time is considered by the client to be of prime importance as this allows the supplier to start work without lengthy negotiation and in the knowledge of what profit level the work will attract. At the other extreme are Fixed Firm Price contracts where the supplier takes all of the risk, as illustrated in Figure 8.1 below. It should be noted however that this type of contract requires a greater level of management and encourages the customer to keep to its original scope of work as any changes will attract a price increase from the supplier.

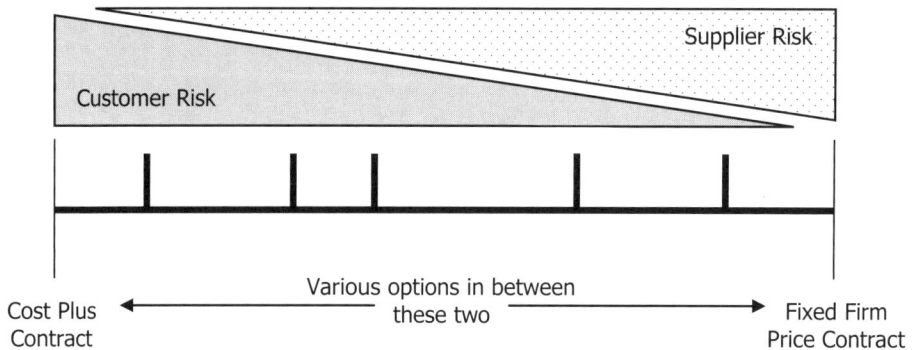

Figure 8.1   Customer-supplier risk allocation

Source: From '*The Essential Management Toolbox*', S.A. Burtonshaw-Gunn (2008). © John Wiley and Sons. Reproduced with permission.

In between these two extremes are a variety of contract types depicted in Figure 8.1. These can cover:

- target cost (expected cost) based on measuring performance;

- target profit (expected profit) which is negotiated as part of the contract;

- Profit Ceiling and Profit Floor arrangement where the maximum and minimum values of the total profit for the supplier is agreed;

- Price Ceiling that the customer will pay and is usually a percentage of target cost;

- sharing arrangement (sharing formula): this is a balance of cost responsibility between the customer and the supplier for each currency unit (£/$) spent;

- point of total assumption. This occurs at the point where the supplier assumes all responsibility for additional costs;

- supplier incentive contracts which are used when the customer wants to motivate the supplier's performance and can attract a greater level of profit (%) if the total costs are reduced or the contractor's performance is improved resulting in programme/time savings, for example. On the other hand the supplier will earn less profit if costs are allowed to increase or the agreed performance targets are not achieved.

In practice all construction contractors will have to include a risk contingency sum in their tender as protection against the risk they have been asked to carry. The interrelationship of risk and reward is illustrated in Figure 8.2 which shows that the requirements are expressed in terms of contractor's incentive, the client's flexibility and the contractor's risk contingency. It is apparent that generally the contractor's incentive and the client's flexibility tend to be incompatible. For example, a lump sum contract imposes maximum incentive on the contractor but also implies a very high level of constraint on the client against introducing change – and hence additional associated costs. The converse is also true at the other extreme of a cost-reimbursable plus percentage fee contract.

Another option, either as an alternative or in addition to risk contingency, is for risk managers to insure against the financial implications of risk occurrence. This can be achieved by payment of a premium to an insurance company who may also choose to spread the risk by sharing this with other insurance companies. The client can either pay directly for this or the contractor can do so though a price arrangement with the client. The risks that are covered by insurance are shown in Figure 8.3.

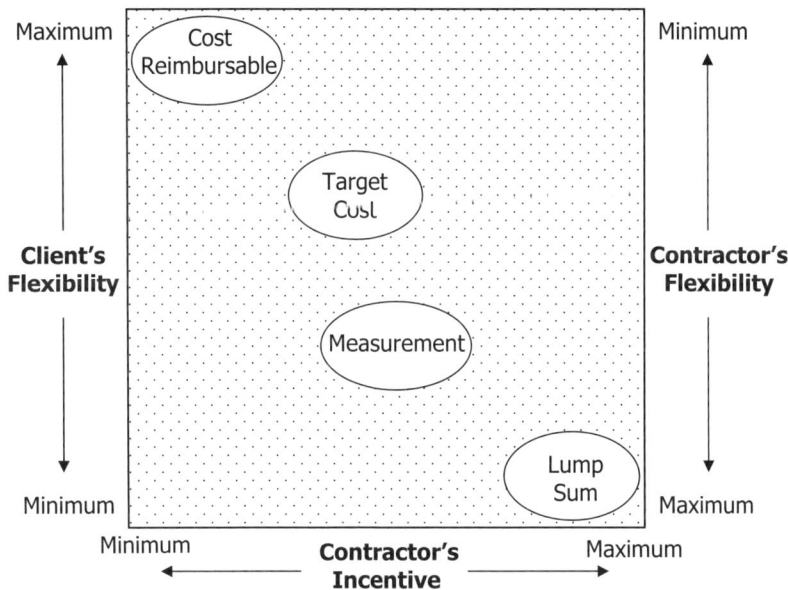

Figure 8.2   Characteristics of different types of construction contracts

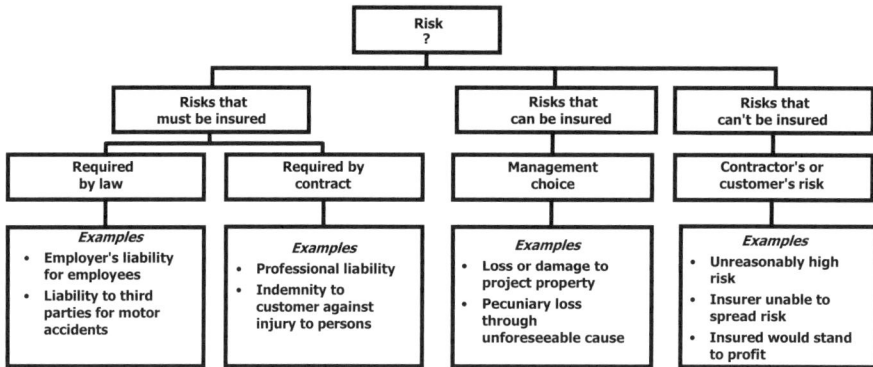

Figure 8.3   Risk insurance options

Source: Model taken from '*Project Management*' 9th edition by Dennis Lock. © Gower
Publications, 2007. Reproduced by kind permission of Dennis Lock and Gower Publishing

Since January 2005 the Financial Services Authority (FSA) regulates
and authorizes all insurance companies and brokers making it illegal for an
individual or company to deal in insurance unless they are FSA regulated.
There are four main classes of insurance:

1. Legal liabilities – payments to others as a result of statutory,
    contractual or professional commitment, compensation awarded
    by the courts, legal expenses but not fines imposed by the courts.

2. Protection against loss or damage to property, including temporary
    works and work in progress, owned construction plant, hired-in
    plant and employee's effects.

3. Cover relating to personnel.

4. Pecuniary loss (that is, involving a money penalty or fine).

An insurance policy may be required for each of these or they may be combined
into a single policy depending on the insurance company chosen. As seen in
Figure 8.3 there are some legal requirements on companies to obtain adequate
insurance cover against some risks, these are those insurance policies which
must be in place in order to comply with laws and regulations, for example,
employers are obliged to insure their employees against injury or illness arising
from their employment (Employees Liability Insurance) and every employer
must display a valid insurance certificate to show that this is in place.

Of particular interest to the construction project manager are the statutory regulations which cover the periodic inspection and certification of lifting equipment, pressure systems and local-exhaust ventilation plant, as no project can include the installation of such equipment without the relevant written scheme of examination and accompanying inspection certificates. If the correct documentation is not supplied the client will not legally be able to operate such equipment. These regulations form part of the UK's Health and Safety at Work Act 1974 and similar requirements apply across all EU member states.

Liability insurances are most likely to feature in construction projects and may be required for:

- compensation to persons for bodily harm (employees or either party, others working on site, visitors and members of the public);

- property loss or damage including work in progress;

- financial loss;

- infringement of property rights;

- accidents;

- product liability (arising from the use of the product);

- professional negligence;

- nuisance caused by works;

- environmental damage.

In addition to the statutory and contractual requirements there is a range of other risks where a contractor may be required to carry suitable insurance. As these relate to the contractor's business this is often a management decision on whether such insurance is considered appropriate or whether the contractor is able to cover any claims from its own funding reserves. These insurance polices include, for example:

- contractor's all risk insurance for construction and engineering projects;

- latent defects insurance;

- accident and sickness insurance;

- key person insurance;

- pecuniary insurance.

There are risks however which insurance companies will either refuse to insure or for which the premium demanded would be very prohibitive and therefore not efficient. Dennis Lock's book on *Project Management* reports these as being:

- Where the chances against a loss occurring are too high or, in other words, where the risk is seen as more of a certainty than reasonable chance. Examples are losses made through speculative trading or because of disadvantageous changes in foreign exchange rates.

- Where the insurer is not able to spread its risk over a sufficient number of similar risks.

- Where the insurer does not have access to sufficient data from the past to be able to quantify the future risk.

- Where the insured would stand to gain as a result of a claim. Except in some forms of personal insurance, the principle of insurance is to attempt to reinstate the insured's position to that which existed before the loss event. A person cannot, for example, expect to benefit personally from a claim for loss or damage to property not belonging to them (property in which they have no insurable interest).

These items must, therefore, be excluded from the insurance portfolio. In some cases other commercial remedies might exist for offsetting the risks. Where accidents are concerned and are covered by insurance policies many companies have produced their own calculations as to the true cost as there are a substantial number of hidden costs, often referred to as the 'iceberg effect' where the uninsured consequential costs may be between 10 and 35 times more than those of the insured injury, ill health and damage costs.

Contract Strategy in Practice

The choice around contract strategy for construction projects in more developed countries includes Design and Build, Turn Key, Project Management and Construction Management. The main advantage of these systems is the speeding up of completion of the investment project thanks to the transfer of some activities to the preparatory phase of the investment. Whilst there have been a choice between these, the construction industry has also seen, through greater supply chain integration, an increased use in the prime contracting (sometimes referred to as PC) system for large-scale projects as discussed in Chapter 5. This approach allows the investor to choose one major contractor (the prime contractor, which is itself usually a major construction company) who in turn is responsible for the designers, the architects, the trade designers and all of the supply chain activities. Very often, this choice is made using tender procedures concluding with the award of a contract and supported by the appropriate contractual agreements.

An advantage of this system for the investor is in the clarity of the agreement concluded with the prime contractor and the remaining participants of the construction process, resulting in a reduction in any conflicts arising from delays in the performance of individual scope of work packages which may negatively influence the completion schedule and the investment cost. In addition any potential errors at the design stage, which can lead to conflict between the designers, the contractors and the investors and can negatively influence the additional costs borne by the investor, are also eliminated, or at least significantly reduced.

One of the factors also affecting the choice of contract strategy is that of the funding required for the project and how the construction elements will be paid for. Typical of high value or long-duration projects is for the contractual arrangement to include payment milestones which attempt to match costs with income to reduce the providers borrowing and hence the overall cost to the client organization. An example milestone cash flow programme is shown in Figure 8.4 illustrating the costs and income of the project funding over time.

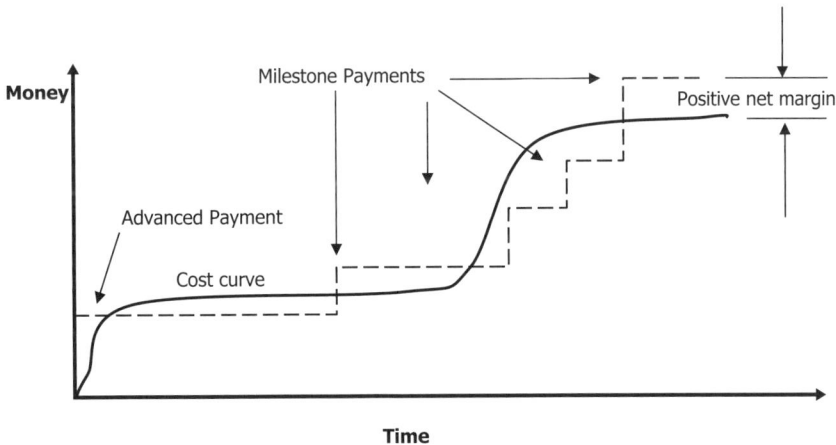

Figure 8.4   Contract cash flow projection – milestone management

Source: From '*The Essential Management Toolbox*', S.A. Burtonshaw-Gunn (2008). © John Wiley and Sons. Reproduced with permission.

The use of the Public-Private Partnership (PPP) contractual arrangement is a growing form of long-term cooperation of the public sector with the private sector in order to undertake a construction project in a more effective manner. Unlike a traditional agreement, the private partner is not remunerated for the construction of the facility, but for the provision of services specified in the license agreement throughout its operations period, for example PPPs are used on social projects such as schools, hospitals and museums.

One of the main benefits for the public sector, resulting from implementation of investments within the framework of PPP, is the ability to finance an undertaking, which would not be feasible on the basis of a traditional agreement requiring public funding. PPP also motivates the private partner to finish the investment on time and to maintain the assumed standards. PPP projects have been found to be more applicable to the following sectors: road and railway infrastructure, public transport, water supply and sewage, airport and port developments, prisons and so on, as the income from these facilities is used to pay the prime contractor a return on its investment over a period of time from the project's operational income.

The main participants of the PPP undertaking are public sector institutions, acting as the client party (government, local government authorities, state agencies and institutions); capital investors cooperating through a PPP company with a public entity; institutions financing the undertaking; subcontractors and other participants, such as advisors, insurance institutions and guaranteeing institutions for example.

In PPP arrangements, some of the risks are transferred to the private sector where they can be managed better than in the public sector. The forms of cooperation between the private and public sector depend on the scope of responsibility transferred to the private sector. The most popular PPP structures are presented as part of Figure 8.5 which shows the development of a traditional project through partnering to PPP contracting where the assets are transferred permanently to the private sector at the end of the contracted project period.

Figure 8.5 does not however include all the models of PPP, as these are still an evolving form of cooperation between the private and the state sectors and specific contractual arrangements are often adjusted for individual projects in order to maximize the economic goals of the project through the most effective risk allocation. Listed below are some of the forms of contracts which are similar to each other in the organizational sense but differ in their

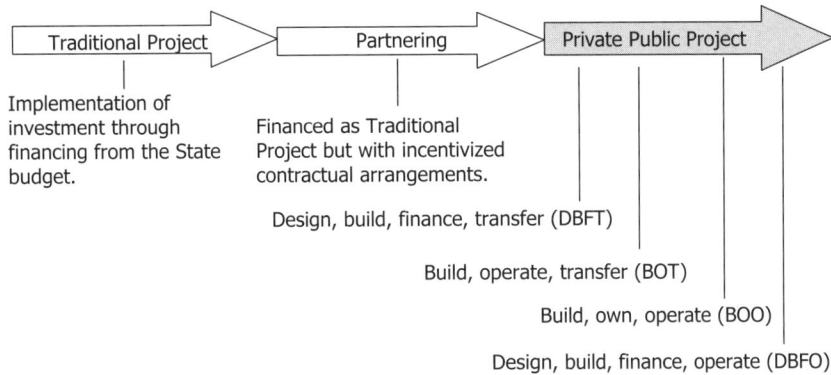

Figure 8.5   Evolution to PPP

financial arrangements and the scope of duties of the parties when used in a
PPP project:

- BOR (build, operate, renew concession);

- BOOT (build, own, operate and transfer);

- BLT or BRT (build, lease or rent, and transfer);

- BT (build and transfer immediately);

- BTO (build, transfer and operate);

- DCMF (design, construct, manage and finance);

- MOT (modernize, own/operate and transfer);

- ROO (renovate, own and operate).

Apart from license agreements, a private entity may also cooperate with a public
one through a joint venture agreement where both entities are shareholders and
are able to share the profits and risks of the implemented investment project.
On the basis of the large number of projects performed within the framework
of PPP, it is possible to identify the advantages and weaknesses of this type of
cooperation. The advantages, together with the weaknesses presented below,
pertain to projects implemented in a mature PPP environment.

The key advantages of this form of contractual arrangement are:

- the speeding up of development of infrastructure;

- the payment for the facility takes place over time;

- the implementation is still possible in the case of limited budget in the public sector;

- the risks are transferred to the parties best able to manage them;

- providing of high quality of services – the supplier is selected on the basis of a tender, and the service level is one of the criteria of selection;

- performance based on requirements; payment depends on the quality and time of delivery of services;

- whole-life costing and the use of a cost management method to optimize costs in the long-term perspective;

- innovations in design, construction and operation of the facility;

- effective financing structures;

- the price competitiveness in the tender process.

On the other hand the main defects of PPP are:

- a very complicated structure compared with the traditional method, requiring the appropriate preparation and abilities of the participants of the project undertaking;

- a lengthened period of preparation of the investment;

- a higher cost of capital for the private sector;

- large tender process costs for the private sector.

Estimating, Budgeting and Cost Control

This final chapter closes with an examination of the financial management of the project both prior to, and during, the main construction phase.

Estimating

In commencing this chapter it has to be said that the first rule of estimating is that there are no magic answers! Furthermore, the second rule of estimating is that poor estimates are everybody's problem! The estimate of project costs is therefore obtained – not by a guess – but from a quantified assessment of the resources required to complete part, or all, of the project. This typical process approach will provide a robust estimate based on previous experience, recorded assumptions and a formula or analogy. Where possible it should be supported by a stated rationale and be achieved by using a range to match any project uncertainty. Additional estimating may then need to be undertaken on areas of the project where wide cost ranges are identified.

Naturally the primary basis for the cost estimate is the project's technical scope of work and as such every element of the project's WBS should be reflected in the cost estimate. If the WBS does not allow the production of an estimate to be made then it is not itself sufficiently detailed or the scope of work is not fully matured on which to produce an accurate price estimate. The technical scope should include information such as:

- a detailed description of the work;

- any work not included in the scope but key to completion – typically this could cover preparation work, materials or even information performed or provided by 'others';

- the WBS elements;

- a description of any regulatory drivers, such as environmental, Health and Safety and construction regulations;

- details of project deliverables, for example, for construction projects this may be the completed project or just the building shell ready for equipment fit-out by other parties.

- any constraints or special conditions on the project. Here a perfect example would be the time constraints for sporting venues to be completed in time for their programmed events or educational facilities which are required in accordance with an academic timetable;

- the Milestone progress dates and associated payments;

- code of accounts;

- project specific considerations such as site access and security for example, both of which will have cost and time implications.

Plainly there are real project management benefits in having accurate estimates of the costs associated with a project, however, the production of a useful estimate takes time, resources and costs money to do well and provide confidence on which to make investment decisions whether at the pre-bid stage or in the execution of the project. It also needs to be recognized that there are a number of common problems in cost estimating, shown in Figure 9.1 together with the factors which contribute to producing a good cost estimate.

Factors which contribute to a good cost estimate	Typical problems encountered in cost estimating
• A reasonably accurate calculation of how much a project might cost. • An evaluation of the cost of resources needed to complete the project. • Estimates that are clear, consistent and assembled in a comprehensive format. • Estimates that are traceable to the technical and schedule elements of the baseline using the WBS coding scheme. • Complete estimates showing all quantities, unit costs and pricing factors for each calculation. • Segregation of costs by major categories – direct, indirect, contingency and escalation for project which run over extended durations. • An estimate that documents all assumptions used to build the estimate.	• The estimate cannot be validated or critically reviewed because it does not detail the assumptions or 'ground rules.' • The estimate does not adequately reflect the baseline schedule, scope or quality specifications. • The estimator's assumptions were too optimistic – and perhaps forgot that not everything works first time! • The estimate does not reflect the construction project in this particular geographic location – costs in one country are not always the same in others. • The estimate contains obsolete factors such as labour rates, overhead costs, failure rates and so on. • The estimator failed to include all of the costs.

Figure 9.1   Contributors and problems of cost estimating

As shown in Figure 9.1, cost estimates should include both direct and indirect costs; these terms are described below.

Direct Costs are those that can be measured and specifically tied to project activities that produce the products or services. Most direct costs are also classed as *variable costs* because the rate at which they are incurred will vary depending on the rate at which work is being undertaken. Typically these include the cost of labour, materials, equipment rentals and subcontractor costs.

On the other hand, Indirect Costs are those which are incurred by the project for common functions that benefit more than one project element or activity and as such cannot be specifically identified with a single project element or activity. Most indirect costs are incurred continuously, at a steady rate: they do not depend on how much work is being performed. Thus they are also referred to as *fixed costs*. Indirect costs are usually assigned to the individual activities on a proportional basis of a function of the direct labour cost: area of space used, for example. Typical categories of indirect costs are overhead, general and administration costs and fringe benefits such as:

- the salaries of the company directors and managers;

- the costs of rent, rates, water, heating and lighting;

- cleaning and maintenance of plant and buildings costs;

- stationery, printing, communications;

- staff salaries in non-direct departments such as accounting, human resources, marketing, office services and security.

The final estimate will therefore include both direct and indirect costs associated with a project and should detail a negotiating margin, the minimum selling price, the profit and the ideal selling price.

Methods of Estimating

It is important to note that the estimating method used needs to reflect the maturity of the project and, as the scope of the work is better defined and the data availability increases, more detailed and reliable estimating methods can, and should, be used. The estimating method used to develop the estimate needs to be known by the project manager because the method will indicate the relative accuracy of the estimate and will reflect how well the project has been defined. In addition, it will also highlight how much data is available on

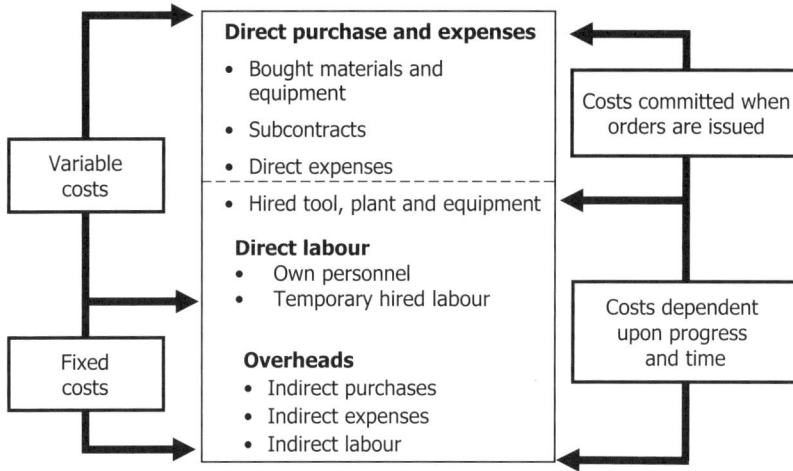

Figure 9.2   Variable and fixed costs

Source: Model taken from '*Project Management*' 9th edition by Dennis Lock. © Gower Publications, 2007. Reproduced by kind permission of Dennis Lock and Gower Publishing.

which to build the cost estimate. Producing an estimate could in practice use more than one method depending on which part of the estimate is better served by a particular approach. There are a number of methods of estimating as listed below:

- analogy;
- parametric;
- expert opinion;
- wideband Delphi;
- range estimating;
- activity based (bottom-up).

Each of these estimating methods is now discussed.

ANALOGY

Analogy estimates utilize statistical analysis including regression analysis to find correlation between costs and performances. An analogy assessment requires assessment of the differences in project elements or technical features and adjustments necessary to accommodate differences, that is, it uses similar projects as a base for the estimate. This can be adjusted for the actual project or actual costs by a factor based upon:

- comparative complexity and design;

- known differences;

- geographical and inflation data.

Analogy estimates need to recognize that the process is assumed but substantially unknown, very little of the technical data or design is available and relies on either reliable estimates or on actual costs from previous projects for comparison.

PARAMETRIC

Parametric estimates rely on the development of Cost Estimating Relationships (CER), based on the characteristics of the project. This relationship is based on an analysis of previously completed projects that are similar to the proposed project in scope, function or materials and so on. This method requires accurate historical data based on similar projects and uses a formula to the driving parameter, for example, cost per square metre, cost per cubic metre of concrete. This requires a meaningful CER to be used with real up-to-date data. For this method to be of maximum value it is important to capture the actual costs from each project that the construction company undertakes in order to update the database as this cost information will improve future CERs and any overall estimates produced in this way.

EXPERT OPINION

Expert opinion can be used to developed estimates where the opinion will say what the object or task will cost. However, this method is only accurate to the extent that the expert is truly up-to-date in both the subject matter and costing from a wide range of experience of the type of project being considered. Expert opinion estimates tend to become more accurate as more experts are consulted, however some caution is required in raising project funds based solely on this approach.

WIDEBAND DELPHI

Wideband Delphi uses the perception of a group of experts to determine an overall consensus estimate. This method combines the expert opinions anonymously in order to avoid 'group think' and any dominant personality from effectively shouting others down. As this requires no face to face contact, the technique works well with email. After the initial briefing on the estimates required and sharing the background materials, the individual experts estimate independently; the coordinator gathers the estimates, collates them and sends them to all of the other expert members for their views. Each expert considers the assumptions, analogies or parameter values and formula used by their peers and either reasserts their original value or incorporates those components of the other people's estimates that they agree with. Using this approach, estimates converge on one or more consensus views each accompanied by the supporting assumptions. The project benefits from understanding which factors are affecting its schedule and cost from all participants. The use of this technique may be restricted to those areas of the project with greatest visibility or political importance, greatest technical or even those areas with third-party risk or the largest impact on dependent activities or groups.

RANGE ESTIMATING

Range estimating is a method which can be used when there is some, but limited, technical data, the scope of work is poorly defined for at least one phase or there is some historical data or project experience on similar undertakings. This method uses a range of estimates to capture the uncertainty in the estimates. The scope of work or the WBS elements are evaluated to determine either:

- the effort required and a degree of confidence (%) for each; or

- a three-point estimate of the most optimistic, the most pessimistic and the most likely estimates for each element, as shown in Figure 9.3. It is important that the estimate also describes or explains the rationale, for example, for explaining the reason for the optimistic estimate and in respect to the pessimistic estimate, what could go wrong.

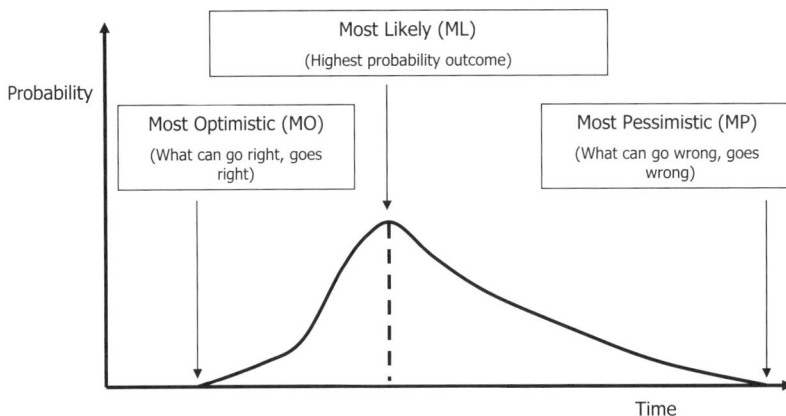

Figure 9.3   Example of three-point estimate for delivery of materials to site

ACTIVITY BASED

Activity-based estimates/activity-based costing (ABC) is commonly called 'bottom-up' costing and is widely used on large-scale projects. Activity-based estimates use a detailed project design to list all materials and labour required to undertaken the project. The cost of materials is estimated at the greatest level of detail and then summed up through the various levels of the WBS. Activities associated with installing material and equipment are estimated using detailed cost estimating relationships for each activity. This method requires the WBS to identify the lowest levels of individually measurable and quantifiable activities on which to build a cost estimate. The project design should be sufficient to establish detailed material take-offs, the methods of performance and any constraints of the activities. It should be noted that the accuracy of the WBS, the design and the historical data directly correlate to the accuracy of the cost estimate using this technique. Use of this technique emphasizes the importance of the end of project review of risk and project information as discussed in Chapter 4.

Budgeting and Cost Management

The two topics of Budgeting and Cost Management are clearly linked; budgeting covers the understanding of what costs will be incurred, when and why, and clearly follows on from the estimating activities and the award of the project. Cost management on the other hand covers knowing what costs have been incurred by the project, when this expenditure happened and what future costs are planned. This is not just a simple administrative monitoring task, as cost management involves ensuring that the sums spent and invoiced are in accordance with the budget, that the timing of each transaction is appropriate and taking necessary action to ensure that any corrective project actions are taken as and when required. In general there are a number of key elements of cost management which need to be produced in addition to good project and subcontract management:

- identified deliverables for a measured outcome;
- realistic budgets and plans;
- a work authorization system;
- an accurate cost collection and reporting system;
- an ongoing review and action process;
- an appropriate change control system.

Managing the cost plan begins with defining and planning the work in a formal and structured way. In addition to the normal steps associated with planning work, the cost plan management will require the development of the measurement methods for the types of work to be performed. This step is critical to assure that the comparisons are valid and reliable. From planning, the process moves to actually utilizing the plan, collecting measurement data, assessing it and taking any necessary corrective actions; this process can be seen to use the same closed-loop feedback approach shown earlier in Chapter 4 (See Figure 4.5). This is the point in the cost management process where the projects cost and schedule performance are obtained and reports generated both for the project team and externally for the stakeholders and customers. One final, but critical, part of the process is the need for continuing maintenance and control of the cost plan. It should be noted that changes to the cost plan can invalidate the measurement process unless these changes are understood and they truly represent a difference from that of the original scope of work.

When assembling a cost plan it will be important to identify when certain costs will accrue. Failing to do this can result in apparent variances of significant sizes, which might be explained simply as cash flow. Whilst the purchase of materials usually impacts the account in one large sum, the cost of a project should ideally be matched to the income as shown in Figure 9.4. In addition actual project cash flow also needs to be considered.

The term 'cash flow' refers to the money that goes in and out of the project. Cash inflow is a positive amount typically from achieving progress or payment milestones on construction activities. Cash outflow by contrast is a negative amount arising from the payment of suppliers, materials, equipment and so on. The difference between the positive cash flow (income) and negative cash flow (outturn) is known as net cash flow, as seen in the table of Figure 9.5.

In general, cash flow at the project level contains a full list of expenses and all amounts earned during the project implementation phase. The items covered in cash outturn are likely to include: the costs of the contract management, cost of preliminary design work, materials and supplies, equipment and rental charges, payment for subcontractors, workers and any additional costs. The cash income which influences the cash flow throughout the project life cycle includes, for instance, the time (delay) in receipt of payments from the client organization, the time of payment on behalf of the subcontractors, crediting conditions, and any equipment rental conditions and so on.

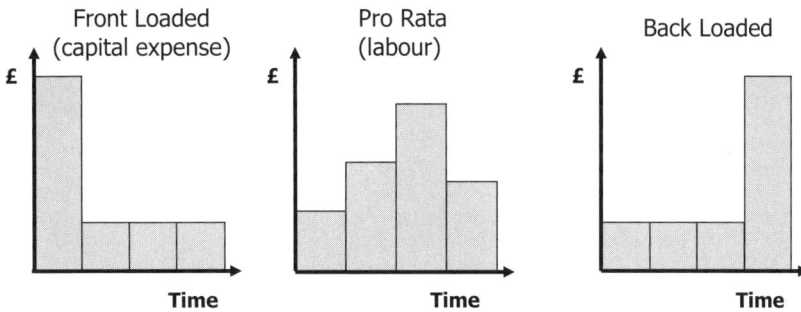

Figure 9.4   Cost profile

Year	0	1	2	3
Income	0	0	5	60
Outturn	40	10	5	0
Net Cash Flow	-40	-10	0	60
Cumulative Cash Flow	-40	-50	-50	10

Figure 9.5   Cash flow

The cost profile is likely to feature as part of the contract negotiations especially on long-term and/or high-value projects and may include provision for payments and income as shown in Figure 9.4.

Many construction projects are described in terms of negative values of cash flow during the construction works and this situation may change after pre-payment of sums made by the client and after the final settlement of the project. The project cash flow is shown in Figure 9.6, the profile of the cash flow for the initial investment, that is, the construction activities would be improved if milestone payments were included as part of the contractual agreement. It should be noted that the cash flow profile in this figure accurately reflects the PPP arrangements where the initial investment is not recovered until typically two-thirds through a project's life, in many PPP projects the break-even point is not reached until 15 years after the construction is complete and the cash recover period begins.

Cost Control

Money is one of the most important resources in business and often prompts the expression that 'Cash is King' – and not least so on construction projects. Many businesses have gone bankrupt not through lack of work but as a result of improper management of its cash flow. In the context of the needs and analysis of the real conditions of the business environment, many methods of cash flow management have been devised. Cost control is therefore concerned with:

- influencing the factors that create changes to the cost plan and ensuring that changes are agreed upon;

- determining that the cost plan has changed;

- managing the actual cost changes as and when they occur;

- monitoring cost performance to detect and understand variances from the cost plan;

- ensuring that all appropriate changes are recorded accurately in the cost plan;

- preventing incorrect, inappropriate or any unauthorized changes from being included in the cost plan;

- taking actions to bring expected costs within acceptable limits.

From the above list, cost control can be seen to include examining and understanding the reasons for both positive and negative cost variances. It is often integrated with other control processes such as scope change control, schedule control, quality control and so on and requires the following tools:

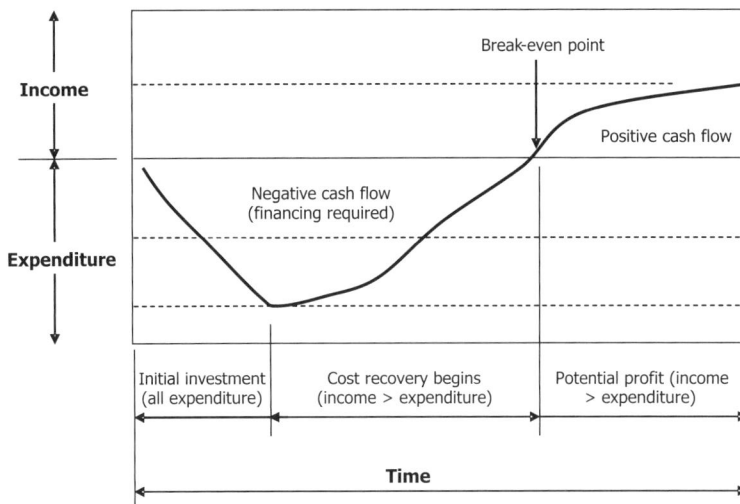

Figure 9.6   Project cash flow

Source: From '*The Essential Management Toolbox*', S A Burtonshaw-Gunn (2008). © John Wiley and Sons. Reproduced with permission.

- a cost plan for the project agreed with the client and key stakeholders;

- performance reports which provide information on the project scope and cost performance such as which budgets have been met and which have not;

- a change request process to capture any changes which occur in many forms: oral and written, direct and indirect, externally or internally initiated, and legally mandated or optional. The changes may require increasing the budget or may even allow it to be reduced in the event of a revised scope of work. A cost change control system defines the procedures by which the cost baseline may be changed and typically includes the paperwork, tracking systems and approval levels necessary for authorizing budget or expenditure changes. The cost change system should also be integrated with the functional or physical change control system.

Tools and Techniques for Cost Control

The available techniques of cost management and forecasting differ with regard to the level of detail and preciseness of the assessment of the actual state and forecasts. In reality, many factors operating during the project lifecycle, such as the time delays, exceeding of costs, changes of orders (and the associated cost changes) influence the real cash flow in the project. The basic problem of forecasting and managing cash flow is the issue of taking into account these factors in order to obtain accurate control. Furthermore, a second challenge is assessing the actual monetary values associated with the project implementation, such as time delays, cost deviations and so on in the context of the work value. Often assimilating all of these factors is very difficult and, sometimes it has to be said, even impossible. Consequently, cost control typically uses detailed plans and schedules which were devised at the early stage of the project life cycle.

Typical 'S' type budget-planned cost curves in Figure 9.7 shows the 'time now' position and the cost value of the work done compared to the budgeted expenditure and the actual costs incurred.

At the 'time now' point, the planned costs are not exceeded by the actual project costs. This position would suggest that while the project is not exceeding the planned rate of spend, it is also not achieving the expected progress for the costs incurred at this time. By conducting a series of regular project reviews

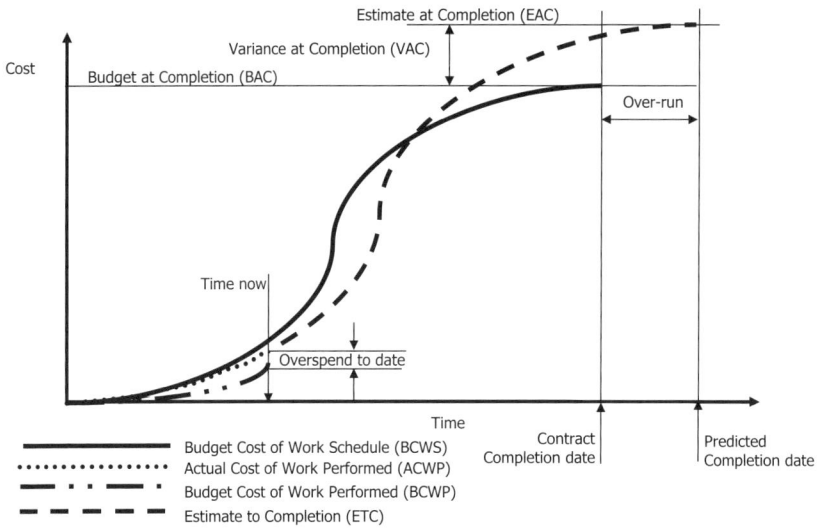

Figure 9.7   Typical cost/time 'S' curve

Source: From '*The Essential Management Toolbox*', S.A. Burtonshaw-Gunn (2008). © John Wiley and Sons. Reproduced with permission.

it is possible to predict the project cost overrun in time and overspend in monetary terms and then examine options to rectify the situation. The variance – the difference between the planned and the actual performance – may be significant and if so will need to be addressed.

The outputs from cost control include the following:

- Revised cost estimates where modifications to the cost information can be used to manage the project. Appropriate stakeholders will need to be notified as appropriate cost estimates may or may not require adjustments to other aspects of the project plan.

- Budget updates which are a special category of revised cost estimates as changes to an approved costs baseline. These are generally revised only in response to changes in the scope of work. In some cases, cost variances may be so severe that re-baselining the project programme is needed to provide a realistic measure of performance.

- Corrective actions are anything done to bring the expected or planned future project performance in line with the desired project plan.

- An Estimate At Completion (EAC) is a forecast of most likely total project costs based on project performance and risk quantification. The most common forecasting techniques are some variation of the following methods:

  - The actual cost to date plus the estimate for all remaining work. This approach is most often used when past performance shows that the original assumptions were fundamentally flawed, or that they are no longer relevant due to a change in conditions. The formula for this is EAC = AC + ETC.

  - Actual cost to date plus remaining budget. This cost calculation is most often used when current variances are seen as atypical and the project management team expectations are that similar variances will not occur in the future. The formula for this is EAC = AC + BAC - EV.

  - EAC = Actual cost to date plus remaining budget modified by a performance factor, often the cumulative CPI. This approach is most often used when current variances are seen as typical of future variances. Finally, the formula here is EAC = (AC + (BAC - EV) / CPI).

Each of these approaches may be the correct approach for any given project and will provide the project management team with a signal if the EAC forecasts go beyond acceptable tolerances. EAC is an extrapolation from current fact and as such it needs to be remembered that this is just an approximation. Care needs to be taken that the results reflect the real project dynamics that exist underneath the 'snapshot' of the cost and expenditure figures. One way to deal with the approximation issue is to compute several EACs to show a range of project outcomes. As covered under Estimating in this chapter, this could be derived from recent experience, expert judgment or the use of various performance factors.

Figure 7.6 in Chapter 7 also lists a number of the above terms and project calculations applicable to construction projects.

References

Association of Project Management (1997) *Project Risk Analysis and Management Guide*, Edited by Simon, Hillson and Newland.

Association of Project Management (2004) *Project Risk Analysis and Management Guide (PRAM)*, 2nd Edition, ISBN 9781903494127.

Banwell, Sir Harold (1964) *Placing and Management of Building and Civil Engineering Work*, HMSO.

Burtonshaw-Gunn, S.A. (2004) Examining Risk and Supply Chain Collaborative Working in the UK Construction Industry, *Supply Chain Risk*, Edited by Professor C. S. Brindley, Ashgate Publishing, ISBN 978-0754639022.

Burtonshaw-Gunn, S.A, (2005) *Considerations of Pre-contract Risks in International PFI Projects*, 2nd International SCRI symposium, University of Salford, April 2005, ISBN 0902896733.

Burtonshaw-Gunn, S.A. (2008) *The Essential Management Toolbox: Models, Tools and Notes for Managers and Consultants*, John Wiley and Sons UK, ISBN 978-0-047051837-3.

Burtonshaw-Gunn, S.A. (2008) The Importance of Pre-Contract Risk Assessment and Management in PFI International Projects, *Supply Chain Risk: A Handbook of Assessment, Management and Performance*, Edited by Professors R L Ritchie and G Zsidisin, Springer International Publications, ISBN 978 0387799339.

Egan, Sir John (1998) Rethinking Construction: The Report of the Construction Task Force to the Deputy Prime Minister, John Prescott, on the scope for improving the quality and efficiency of UK construction, © Crown Copyright 1998 URN 03/951.

Emmerson, Sir Harold (1962) *Survey of Problems Before the Construction Industries*, HMSO.

Industry Guidelines on a Framework for Risk Related Decision Support, UK Offshore Operations Association (UKOOA), Issue 1 May 1999, Industry publication reference EHS08.

Latham, Sir Michael (1994) *Constructing the Team: Joint Review of Procurement and Contractual Arrangements in the United Kingdom Construction Industry*, HMSO, ISBN 011-752994X.

Lock, D. (2007) *Project Management*, 9th Edition, Gower Publishing Limited, ISBN 978-0566087691.

National Audit Office (2003) *PFI: Construction Performance*, Report by the Comptoller and Auditor General HC371 Session 2002–2003: 5 February 2003, The Stationery Office.

Porter, M.E. (1985) *Competitive Advantage: Creating and Sustaining Superior Performance, The Free Press*, a Division of Simon & Schuster Adult Publishing Group Free Press, ISBN 978-0684841465.

Project Management Institute (1992) *Project & Program Risk Management: A Guide to Managing Project Risks and Opportunities*.

Rail Safety & Standards Board (2004) *Decision-making Practices and Lessons from Other Industries*, Report T266.

Simon, Sir Ernest (1944) *The Placing and Management of Building Contracts*, HMSO, London.

Smith, N., Merna, T. and Jobling, P. (2006) *Managing Risk in Construction Projects*, Blackwell Publishing, ISBN 140-5130121.

Further Reading

Association of Project Management (2006) *Body of Knowledge*, 5th Edition, ISBN 9781903494134.

Barnes, M. (1985) *Project Management Framework, International Project Management Yearbook*, Butterworth Scientific.

BS 6079-3:2000 Project Management (2000) *Guide to the Management of Business Related Project Risk*, ISBN 0580331229.

BS EN 12973:2000 Value Management, ISBN 0580-35686-8.

Edwards, P. and Bowen, P. (2004) *Risk Management in Project Organisations*, Butterworth-Heinemann, ISBN 978-0750666299.

Institution of Civil Engineers (ICE) (2005) *Risk Analysis and Management For Projects: A Strategic Framework For Managing Project Risk and its Financial Implications*, 2nd Edition.

Jones, M.E. and Sutherland, G. (1999) *Implementing Turnbull – A Boardroom Briefing*, Published by the Institute of Chartered Accountants in England and Wales.

Kelly, J., Male, S. and Grahan, D. (2004) *Value Management of Construction Projects*, Wiley Blackwell, ISBN 978-0632051434.

Loosemore, M. *et al*, (2005) *Risk Management in Projects*, 2nd Edition, Taylor and Francis, ISBN 978-0415260566.

Shutt, R. C. (1995) *Economics of the Construction Industry*, 3rd Edition, Longman, ISBN 0582-22912.

Useful Websites

UK Health and Safety Executive, www.hse.gov.uk/risk
Institute of Risk Management, www.theirm.org/index.html

Glossary

Accrual	Work done for which payment is due but has not been made.
Action plan	A strategy for the mitigation of a risk.
Analysis	The qualitative component of the process in which the identified risks are considered to determine their importance, possible effects and required actions.
Business risks	Those risks that affect the operation of the business outcome once it has been delivered by the project.
Construction	The activity of assembling materials and components designed and produced by a multitude of suppliers, working in a diversity of disciplines and technologies, in order to create the built environment.
Commitment	A binding financial obligation, typically in the form of a purchase order or contract.
Credit risk management	The risk of failure of a third party to meet its contractual obligations to the organization under a project arrangement particularly as a result of the third party's diminished creditworthiness and the resulting detrimental effect on the organization's capital or revenue resources.
Direct costs	These costs can be measured and are specifically tied to project activities that produce the products or services.

Environmental risks	Risks that are external to the project environment but which nevertheless can affect the project objectives. For example, the Gulf War had a devastating effect upon gas field projects in Kuwait in 1990.
Evaluation	The quantitative component of the process in which effects are evaluated to generate action.
External change risks	Risks that are beyond the immediate project environment but which could have a major impact. Frequently in contractual terms these may include *force majeure* events. However, external change risks go beyond this, for example, because of a change in government policy or in its interpretation of a law.
Exchange rate risk management	Exchange rate risk management is the risk that fluctuations in foreign exchange rates have an adverse effect on the organization's finances, against which the organization has failed to protect itself adequately or a favourable effect has been missed. It should be noted that organizations should try to avoid unnecessary exposure to exchange rate fluctuations.
Fallback plan	A plan of a strategy in the case where an action to a risk in unresponsive.
Hazard	A situation with a potential to cause harm to people or the environment or affect the objectives of a task.
Indirect costs	Indirect costs are those which are incurred by the project for common functions that benefit more than one project element or activity and as such cannot be specifically identified with a single project element or activity.
Inflation risk management	The risk that prevailing levels of inflation have an adverse effect on the organization and project against which the organization has failed to protect itself adequately or a favourable effect has been missed.

Interest rate risk management	The risk that fluctuations in the levels of interest rates have an adverse effect on the businesses finances, against which the organization has failed to protect itself adequately, or a favourable effect is missed. It should be noted that organizations will need to manage their exposure to fluctuations in interest rates with a view to containing interest cost or securing its interest revenues while maintaining the security of the invested funds. They may achieve this by the prudent use of investments to create stability and certainty of costs and revenues.
Invoice	A demand for payment of goods, materials or services received.
Impact	The measure of an effect expressed quantitatively in the affected domain, for example, when a project is delayed by one month or its cost increases by £10 000.
Legal and regulatory risk	The risk that the organization itself, or a third party with which it is dealing in its treasury management activities, fails to act in accordance with it legal powers or regulatory requirements and that the organization suffer loss accordingly. All organizations should ensure that all their treasury management activities comply with statutory powers and regulatory requirements.
Liquidity risk management	This is defined by the Chartered Institute of Public Finance and Accountancy (CIPFA) Code of Practice as the risk that cash will not be available when it is needed, that ineffective management of liquidity creates additional unbudgeted costs and that the organization's business objectives will therefore be compromised. To counter this an organization will need to ensure that it has adequate, though not excessive, cash resources, borrowing arrangements, overdraft or standby facilities to enable it at all times to have the level of funds available which are necessary for the achievement of its business objectives.

Market risk management	This risk arises from adverse market fluctuations in the valuation of the principal sums an organization invests and that its stated treasury management policies and objectives have failed to protect it adequately.

It should be noted that organizations should seek to ensure that their stated treasury management policies and objectives are not compromised by adverse market fluctuations in the value of the principal sums it invests. |
Monte Carlo simulation	Monte Carlo simulation was named after Monte Carlo, Monaco, where the primary attractions are casinos containing games of chance which exhibit random behaviour. Monte Carlo simulation selects variable values at random to simulate a risk model.
Payment	Monetary return for services rendered or goods/ materials received – discharge of indebtedness.
PESTLE	An analysis technique used to gain an assessment of the project or situation from a Political, Economic, Social, Technology, Legal and Environmental perspective.
Post mitigation	A term used to relate to impacts, probabilities, criticality score and exposure as assessed with account taken of commitments to mitigation actions, including an assessment of their likely success and effectiveness.
Probability of occurrence	A judgement of the chance that an identified risk will materialize. Probability may be measured and presented in bands (for example, high, medium or low) or as a percentage or a decimal number.

Project risks	Risks within the project boundary that could affect the delivery of the business outcome that the project is set up to deliver. In other words, those that could affect the delivery of the project's time, cost and specification objectives.
Residual risk	The risk remaining after risk control measures have been implemented.
Reporting	The action of the risk process which links analysis, evaluation and mitigation to inform the client, project team and stakeholders.
Risk	An expression of the potential of a hazard to cause harm to people or the environment or affect the objectives of a task.
Risk assessment	The process of measuring risk, where risk is a function of probability and consequence.
Risk control	Measures taken to eliminate, reduce or protect against risk.
Refinancing risk management	The risk that maturing borrowings, capital, project or partnership financings cannot be refinanced on terms that reflect the provisions made by the business for those refinancing, both capital and revenue and/or that the terms are inconsistent with prevailing market conditions at the time. Organizations should ensure that borrowings, private financing and partnership arrangements are negotiated, structured and documented, and the maturity profile of monies so raised are managed with a view to obtaining offer terms for renewal or refinancing if required which are as competitive and favourable as can be reasonably achieved at that time.
Secondary risk	A risk associated with a fallback plan.
Terminate	Pre-emptive risk response action of termination or avoidance centres on changing the project plan to eliminate the risk or to protect the project objectives from its impact.

Tolerate

This risk strategy indicates that the project has decided not to change the project plan and to deal with a risk or is unable to identify any other suitable strategy to adopt.

Transfer

Risk transfer is seeking to move the consequence of a risk to a third party together with ownership of the response. Transferring the risk does not eliminate it; it simply gives another party responsibility for its management. This is the most effective way of dealing with financial risk exposure and can be by a contract to another party or by payment of a premium in the case of insurance.

Treat

This strategy seeks to reduce the risk probability or its impact by taking early action to reduce the occurrence of the risk to an acceptable limit. It may take the form of implementing new processes, undertaking more preliminary work or selecting more stable suppliers.

Uncertainty

A condition that may or may not occur, most commonly associated with novelty or advances in technology, or process or resources stretch, or where interfaces with third parties can influence project success.

Value Analysis (VA)

This is the title given to value techniques applied retrospectively to completed projects to 'analyze' or audit the project's performance

Value Engineering (VE)

This is the title given to value techniques applied during the design or 'engineering' phases of a project. Thus, VE studies are those conducted during the early part of the working drawings stage.

Value Management (VM)	This is the title given to the full range of value techniques available. It is a higher order title and is not linked to a particular project stage at which value techniques may be applied. It maximizes the functional value of a project by managing its evolution and development from concept to completion, through the comparison and audit of all decisions against a value system determined by the client or customer.
Value Planning (VP)	This is the title given to value techniques applied during the 'planning' phases of a project. Thus, VP studies are those conducted during the briefing or sketch plan stages.

About the Author

Professor Simon A. Burtonshaw-Gunn is a Principal Management Consultant with Risktec Solutions Limited: a specialist risk management consultancy company predominantly involved in the high-risk and externally-regulated industries. He is well qualified with an MA degree in Strategic HRM, an MSc in Business Management and a PhD from research into construction industry partnering and collaborative working under the supervision of Dr Bob Ritchie, now Professor of Risk Management at the University of Central Lancashire. To complement his academic achievements his experience is recognized with Fellowship of four UK professional institutions including, for over 10 years, the Association of Project Management. He has over 30 years experience across a number of sectors, most notably in high-technology where his activities have included a wide range of management system developments, risk management, project management, the management of change and people-focused performance improvements in both private and public sectors.

Covering over 400 000 miles travel by air he has undertaken specialist consultancy work in Russia, Belarus, Tatarstan, the Ukraine, Libya, the Republic of Kazakhstan, Indonesia, the Kingdom of Saudi Arabia, Qatar and the UK. In addition he has designed and delivered management training as part of a European Union cooperation programme in Bangladesh, Thailand, Indonesia, Sri Lanka, Malaysia, the Philippines, Vietnam and India.

For a period of 4 years he held the position of a Post-doctorate Research Fellow at the Manchester Metropolitan University relinquishing this at the beginning of 2005 to take up the role of a Visiting Professor at the University of Salford in Greater Manchester. With a strong interest in education, Professor Burtonshaw-Gunn has been a research examiner for the UK's Chartered Institute of Purchasing and Supply (CIPS) since 2002 and has been a member of the Resources Committee of the Board of Governors at the University of Bolton between 2006 and 2008. Since 2007 he has been a member of the Court at the University of Leeds under the direction of the University's Chancellor Lord (Melvyn) Bragg.

Simon is one of the founding members of the academic research group ISCRiM (International Supply Chain Risk Management) and has contributed chapters to the two books that this group has published to date in addition to presenting conference papers in Sweden, the USA, UK, Hungary and India on risk-related topics. Professor Burtonshaw-Gunn has written over 40 refereed publications, professional journal articles, textbook contributions and conference presentations. His first book specifically for managers and management consultants '*The Essential Management Toolbox*' was published in 2008.

Index

4Ts 30–1, 74–5, 77, 109, 124

Action planning 7, 63
Activity based costing (ABC) 157
ALARP 77–9
Analogy assessment 154
Association of Project Management 84
Assumptions analysis 42

Bank loans 89, 121
Banwell Sir Harold 5
Bottom-up risk identification 39
Bow-tie 13, 66–7
Brainstorming 42–3
British Standards 9, 130
Budgeting 116, 157
Burtonshaw-Gunn Simon A. 12, 31, 34, 51, 64, 75, 81, 91–2, 97, 132, 142, 147, 160, 162
Business Risk 37

Cardinal scales 53–4
Cash Flow 147, 158–62
Checklists 42, 55–6
Competition 4, 108
Consequences 9, 12–3, 24, 25, 29, 32, 41, 60–3, 65–6, 69, 81, 85, 103, 124
Constructing the Team 5, 6, 90
Construction activity 1
Construction companies 1–2, 4, 30, 90
Construction contracts 143
Construction industry 1–3
Construction management 2, 146
Construction projects 7, 8, 11, 14, 24, 29, 32–3, 44, 54, 89, 97, 103, 114, 125, 128–9, 145–6, 152, 159, 163
Contingency 9, 14, 35, 49–50, 63, 65, 68–70, 73, 75, 81–2, 93, 109, 134, 140, 143, 152

Contract strategy 15, 109, 137–38, 141, 146–7
Contract types 99, 138, 140, 142–3
Cost control 11, 15, 27, 116, 159–62
Cost estimate 15–6, 39, 109, 134, 151–4, 157, 162
Cost Estimating Relationships (CER) 155, 157
Cost plan 158, 161
Customer-Supplier risk 142

Delphi techniques 43, 154–5
Direct costs 153
Discounted cash flow 126–7
Document review 42
Dragon's Den 116

Earned value 41, 131–3, 135–6
Egan, Sir John 6, 89–90
Emmerson, Sir Harold 5
Environmental risk 37
Estimating process 16, 151–3

Factoring 119
Fallback Plan 49, 68, 76, 81–2
Feedback 35, 77, 81–2, 158
Fishbone diagram 42, 61, 64
FMEA 61, 64, 67
FMECA 67–8
Foreign bonds 118
Funding 114

Governance 12, 22–4, 97
Government 2–6, 37, 40, 50, 89, 94–9, 118, 148

Health and Safety 2–4, 30, 40, 78, 145, 151
High-risk projects 1, 77

Impact definitions 52, 53
Indirect costs 153
Institution of Chartered Accountants 22
Insurance 4, 11, 15, 27, 31, 75, 97, 143–6,
    148
Internal rate of return 117, 128
International projects 104
International risks 97, 99
Interviewing 43
Investment decisions 11, 104–5, 108, 110,
    126–7, 152
Investors 15, 105, 110–3, 127–8, 130

Joint venture 149

Latham, Sir Michael 5–6, 89–90
Learning from experience 44, 84
Leasing 94, 119–20
Lock, Dennis 9, 144, 146, 154
Life-cycle of construction project 28

Monte Carlo simulation 65, 69–70
Mortgage 116

National Audit Office 89, 91
Net Present Value 127
Normal distribution 71, 113

Oil and gas industry 13, 66
Operational risks 28, 93, 95, 98
Ordinate scales 53, 54

Parametric estimates 154–55
Partnering 6, 89–90, 92–3, 148
Payback assessment 126
Performance-Based Management 131
PERT 70
PESTLE 95, 140
Porter, Michael 2, 5
Post project reviews 83, 87
Pre-contract risk management 28, 92, 97
Pre-investment phase 103–5, 109
Prime Contracting 14, 89, 92–5, 100, 146
Prime contractor 28, 90–3, 95–9, 147
PRINCE 84–5
Private Finance Initiative (PFI) 90–2, 94–5,
    97–100
Probability 24, 51–3, 61, 74
Probability definitions 52

Probability impact matrix 53–4, 61
Procurement 4, 24, 39, 40, 89, 92–3, 100,
    112
Project closure 8, 83–6
Project funding 14, 104–6, 114, 117, 135,
    147
Project Management 8, 21–2, 35, 39, 41, 44,
    69, 73, 83, 93, 146, 152
Project Manager 7, 8, 12, 21, 24, 38, 43, 47,
    50, 69–70, 75, 81–2, 85, 134–5, 137,
    139–40, 145, 153
Project performance 7, 14, 68, 123–5, 131–2,
    134, 162
Project plan 15, 21, 35, 42, 44, 47–51, 59, 75,
    79, 85, 93, 109, 132, 162
Project risks 37, 40–1
Project 'S' Curve 132, 161–2
Project teams 21, 22, 24, 38, 42, 54
Project valuation 110, 111
Public-Private Partnership (PPP) 15, 89,
    90–2, 104–5, 118, 148–50, 159

Qualitative risk analysis 13, 46–8, 59, 60–1,
    65
Quantitative risk evaluation 13, 46–8, 59,
    65

Rethinking Construction 90
Return on Investment (ROI) 127
Risk actions 24, 31, 79
Risk control 12–3, 25, 35, 65, 73, 81
Risk decision making 30, 113
Risk driven project management 8, 29
Risk identification 13, 25, 38–9, 55–6, 65
Risk impact 51–3, 74, 78
Risk management steps 24–5, 29
Risk matrix 61–3
Risk mitigation 11, 25, 38, 49, 63, 73
Risk monitoring 14, 79, 80–3
Risk plans 13, 21, 25, 35, 44–7, 74, 82
Risk register 48–9, 68–9, 113
Risk response 14, 25, 30–1, 39, 46–7, 49, 59,
    73–5, 80–1, 93
Risk reviews 30, 49, 73, 77
Risk-reward balance 12, 30, 143
Roles and responsibilities 45–6, 86–7

Simon, Sir Ernest 5
Smith, Nigel 109, 110

Stock Exchange 11, 22, 112, 114
Strategic objectives 3, 137
Suitably Qualified and experienced Personnel (SQEP) 24
Supply Chain Management 90–3, 95–9, 104, 137, 146
SWOT analysis 43

Three point estimates 70–1, 76, 156
Top down risk identification 39

Turnbull Nigel 22

UKOOA 32, 38

Value chain 1
Value Engineering 93
Value Management 15, 93, 128–30
Variance 132–6, 158, 160, 162–3
Venture capital 115–6

If you have found this book useful you may be interested in other titles from Gower

Design for Sustainability
A Practical Approach
Tracy Bhamra and Vicky Lofthouse
Hardback: 978-0-566-08704-2
e-book: 978-0-7546-8775-7

Health and Safety in Construction Design
Brian Thorpe
Hardback: 978-0-566-08670-0

Improving People Performance in Construction
David Cooper
Hardback: 978-0-566-08617-5

International Construction Contract Management:
A Compendium of Knowledge
Bryan Morgan
Hardback: 978-0-566-08697-7

Project Management 9
Dennis Lock
Hardback: 978-0-566-08769-1
Paperback: 978-0-566-08772-1
e-book: 978-0-7546-8634-7

GOWER

Project Management in Construction
Dennis Lock
Hardback: 978-0-566-08612-0
e-book: 978-0-7546-8307-0

Quality Management in Construction
Brian Thorpe and Peter Sumner
Hardback: 978-0-566-08614-4

Understanding and Managing Risk Attitude
David Hillson and Ruth Murray-Webster
Hardback: 978-0-566-08798-1

Using Earned Value
A Project Manager's Guide
Alan Webb
Hardback: 978-0-566-08533-8

Winning New Business in Construction
Terry Gillen
Hardback: 978-0-566-08615-1
e-book: 978-0-7546-8549-4

Visit **www.gowerpublishing.com** and

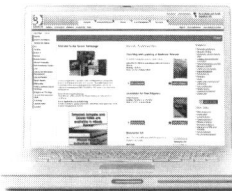

- search the entire catalogue of Gower books in print
- order titles online at 10% discount
- take advantage of special offers
- sign up for our monthly e-mail update service
- download free sample chapters from all recent titles
- download or order our catalogue